Library of Congress Cataloging-in-Publication Data

Kinney, Jane
  Careers for environmental types and others who respect the earth /
Jane Kinney, Michael Fasulo
    p. cm.
  Includes bibliographical references.
  ISBN 0-8442-4102-4 — ISBN 0-8442-4103-2 (pbk.)
  1. Environmental sciences — Vocational guidance — United States.
I. Fasulo, Michael. II. Title.
GE60.F37 1993
363.7'0023—dc20   92-47379
                                                   CIP

1996 Printing

Published by VGM Career Horizons, a division of NTC Publishing Group.
© 1993 by NTC Publishing Group, 4255 West Touhy Avenue,
Lincolnwood (Chicago), Illinois 60646-1975 U.S.A.

5 6 7 8 9 0 VP 9 8 7 6 5 4 3

About the Authors

Michael Fasulo has a master's of arts degree in sociology from The Pennsylvania State University where he now teaches in the Department of Sociology. His areas of expertise include trends in public opinion toward the environment, national and international environmental policy, and environmental risk perception. Michael lives in New Jersey.

Jane Kinney lives and works in New York City. An acquisitions editor for HarperCollins publishers college division, Jane is responsible for signing and developing new titles annually for the publisher's English list. Holding a master's of arts degree in Italian literature, Jane has previously taught Italian at the University of Wisconsin, Madison.

Contents

Foreword

I grew up on the most polluted lake in the country. At that
time—over 30 years ago—the words environment and environmentalist had not yet entered our vocabulary in any significant way. I had no idea how the lake got that way, what it
meant, or who could help solve the problem. All I knew was that
I couldn't swim in the lake, and that seemed very wrong.

Now, of course, an environmental ideology permeates our
society. There are tens of thousands of community groups ad-
dressing local pollution problems, and many other national and
international groups advocating environmental policy. Even
some corporations, long seen as the perpetrators of much of the
damage, are positioning themselves as caretakers of the environment. Yes, the term "environmentalist" is a vague one, subject
to much interpretation and different levels of sincerity. Funda-
mentally, however, all of us want to do what we can to ensure a
healthy environment for all life on the planet.

In my 14 years as an environmental activist, the same questions came forth: What can I do? Where can I find work in
environmental protection? Or in community organizing? Or in
governmental environmental policy, corporate acountability,
the environmental sciences? For too long, my answers were
neither extensive nor systematic. So, it is a welcome sign that

there are now guides to environmental careers, resources where people can help connect their idealism with employment opportunities.

Today, as our society comes under increasing environmental pressures and risks, all of us, but especially young people, should have the opportunity for jobs whose essence includes the right, if not the duty, to take our conscience to work every day. As you use this resource, keep in mind that this is a new and developing field. What you will bring to these jobs will in no small part help create our future. Be careful in ensuring that the values of the places you may work are compatible with your own as a human being, and that you will be allowed to use not only your mind, but also your heart.

Peter Bahouth
Greenpeace

ONE

Career Opportunities for the Ecologically Minded

Our intent in writing this book is to provide you, the reader and concerned environmentalist, with a broad overview of career options in one of the most interesting and dynamic job areas today—the environmental field. This career field is relatively new, in fact its beginnings can be traced directly to the first Earth Day in 1970, and its growth closely parallels the increasing public concern with the state of our fragile environment.

Throughout the book we will give you an essence of the many career paths that you can take. We will discuss job opportunities in corporate America, the government, the media, environmental organizations, and the rewards that await the spirited entrepreneur. All this will be prefaced with an explanation of the value and importance of relevant college education programs. We believe that this broad cross section of career fields, the jobs that fall within each sector, and their particular educational requirements will provide you with a clear and accurate picture of the evolving environmental career landscape. Our main criteria for choosing these career fields were their overall job growth potential and the importance of their contribution to the health of the environment.

In most instances we not only describe job opportunities but also provide you with numerous references to other, more detailed information sources. We have included in our scope not only students but also those already working who are considering a mid-career change. No matter if you are a high school or college student, or a secretary, lawyer, accountant, or sales agent, there are career paths discussed in this book that match your skills to a job that fulfills your desire to work for a better environment.

Most environmental career guides written in the last few years talk almost exclusively of technical and scientific jobs located in private industry and government. While this job pool is indeed large and discussed throughout this book, we believe that the shortfall of these other guides is their failure to address the influence that the environmental movement has had on the entire American job scene. Americans are demanding more environmental sensitivity from both government and the business community and this has provided increasing opportunities for individuals from every conceivable walk of life to put their skills to work for the environment.

No one book can describe every environmentally focused career or all the components for a successful job search, but a well-written and useful career guide will give the reader a better sense of direction and a greater degree of certainty in considering career options. We firmly believe our book will do just this for you!

A Glance at the Environmental Job Horizon

The job opportunities in the environmental field have evolved from their beginnings in the 1960s when career paths centered exclusively around the conservation sciences like natural resource and wildlife management. Today, there is a much greater need for environmental professionals in all sorts of capacities.

For those with an interest in the physical sciences, the three R's: remediation (the cleaning up of toxic pollution), recycling (developing collection systems and product machinery), and restoration (managing areas that are recovering from ecological damage) will provide thousands of new jobs annually, well into the next century. Technicians, who outnumber professionals by three to one in most science-oriented fields, are in the highest demand. These are the individuals who bridge the gap between the theoretical knowledge of an engineer and the skill of a precision mechanic. Technicians are an integral part of any environmental project because they do a majority of the actual work.

Those with a strong knowledge of the social sciences are also in demand. It is only with the help of people who have sound communication, organizational, and management skills—the strength of college liberal arts majors—that the task of repairing and preserving the environment will be made a reality. Policy analysts, educators, fundraisers, and market researchers are just a few of the types of professionals in demand.

In 1990, the EPA estimated that cutting pollution to meet federal standards was a $115 billion venture. In 1985 the market for environmental services and products was a $70 billion business, while today that number is approaching $100 billion. Thus, there are also many opportunities waiting for those who have business savvy with an environmental twist.

Making A Career of Saving the Earth

What does it take to be an environmentalist? Or to pose the question more personally, what can you do throughout your life to make our earth a cleaner and better place to live? If you feel an intense need to do something about the problems with our environment, and you are just starting down a career path or perhaps thinking about making a career change, why not consider making a career out of your environmental concerns! Haven't you always been told that the most satisfying career is the one in which you are most interested? When you look around at your physical surroundings and hear the daily reports of increasing water and air pollution, the growing hole in our ozone layer, and the destruction of our rainforests, do you feel a sense of urgency deep inside? Do you feel that the government and business is paying too little attention to environmental problems and, as a consequence, far less than necessary is being done to solve these problems? Out of concern and perhaps frustration, do you feel the need to take an active part in making our environment a better and healthier place to live?

If you agree with any of these statements, you may be one of the growing number of individuals who want to make a career out of your concern for the environment. Up to this point you may have had an intense desire to do something positive for the earth, but have probably lacked the information to explore your career options.

Unlike other, more conventional career options, there is precious little written about careers for the ecologically minded. In addition, few people seem able to give advice about environmental career options because they are simply not familiar with the many different types of environmental jobs. These job hunting limitations are much less a reflection of the environmental job possibilities—for there are many—but more an example of how recently many environmental job positions have come into being. After reading this book and becoming familiar with some of the resources that we have provided, you will have a solid idea of how to match your interests and personal goals with a specific environmental career path. You will also be able to give others a better understanding of the scope of environmental careers and, in the process, convince them to join you in the fight to save our fragile earth!

Who Are These Career-Minded Environmentalists?

In the last 20 years, great strides have been made to restore and preserve our environment. In the early 1970s, soon after the activities of the first Earth Day, the American public began to show a great interest in the preservation of our physical world. People finally began to realize that for all the accomplishments that humankind could claim, we have also done great and potentially irreversible damage to the earth. It suddenly became apparent that our air, water, and soil were quickly becoming poisoned, and that we were just beginning to feel the devastating effects. The battle cry was sounded and the public responded with great resolve. Since the eve of the environmental movement, public concern for the environment has continued to soar.

As a direct result of this intense public concern and pressure, the President and Congress formed the United States Environmental Protection Agency in the early 1970s and since that time, virtually every act of pro-environmental legislation has been enacted. These facts show that the concerns and actions of ordinary individuals can and will make a difference. The people who began the fight for the environment are still going strong, and many more individuals have joined in. As a result of these activities, the number and types of environmental jobs have also grown tremendously.

We all probably have a common misconception of the people we would identify as typical working environmentalists. These individuals tend to be young, carefree, and willing to do just about anything to make ends meet. When one thinks of the types of jobs that these people have, one immediately conjures the image of an environmental activist, with clipboard in hand, going door to door trying to solicit donations. While this type of work offers the ecologically minded canvasser great grassroots experience, it can also be a lonely and disheartening job, because financial rewards are limited and there are often few career opportunities within any organization. Although this type of work is very important, and often is a stepping stone for the career-minded environmentalist, it is but the tip of the iceberg in terms of the potential pool of environmental careers.

Our stereotypical environmentalist is only one of the many players in the environmental field. In the past twenty years, there has been an explosion of occupations concerned with the environment. Along with and in response to growing public concern about the environment, a whole host of jobs in both the private and public sector have been created or greatly expanded. In this field there is room for everyone; from those who seek a "conventional job" and wish to climb the corporate ladder to occupational prestige and job security, to those who prefer jeans and sneakers and the autonomy of being able to work independently. A number of these jobs take place in the structured office environment while many others are conducted in the outdoors, the courtroom, in a lecture hall, or on the floor of your own business.

What we hope to convey to you is the sense that the career world is full of jobs for the ecologically minded. Further, the rewards for choosing an environmental career are quite appealing. Not only will you be doing something personally satisfying, but the contribution that you make through any of the following environmental career paths will be of net benefit to the earth. It is only through the work of concerned people like you that we will ensure the continued health of our planet and just as importantly, ensure a bright and healthy future for our children and the children of future generations.

A Quick Look at Jobs for the Ecologically Minded

Today there are environmental professionals working in corporations, in all levels of government, in industry as engineers, scientists, and technicians, in the various media, in nonprofit organizations, and as educators and entrepreneurs. Many environmental professionals have scientific and technical backgrounds, although an increasing number have been trained in the liberal arts.

While a major portion of these jobs are in the private sector, there are many opportunities in the public and nonprofit sectors as well. The salaries and growth opportunities are as diverse as the types of ecological jobs. Depending on the size and health of the company and economy, the private sector has traditionally been the place with the greatest earning and growth potential. Conversely, the advantage of a public sector career is job security and job-related benefits. Also, both state and federal agencies create and enforce environmental regulations, which along with being a rewarding experience in itself, can also lead to later job opportunities in the private sector. Nonprofit organizations have traditionally lagged in both job security and salaries but that is all changing. During the last ten years the nonprofit sector has enjoyed tremendous growth, and there are now thousands of environmental organizations nationwide with a growing corps of salaried workers and hundreds of thousands of volunteers. This area is very dynamic because it is where all the grassroots organizing and community action occurs. As you are beginning to see, there is truly a place for everyone who is interested in an environmental career.

Is An Environmental Career Really for You?

One of the first steps in making a major decision like choosing or changing a career path is to examine your motives and assess your qualifications. Some questions you may want to ask yourself are as follows.

1. Do I consider myself already to be an environmentalist? Is my life-style consistent with my environmental beliefs?

2. What activities have I participated in that are related to the environment?

3. Is the integration of my values and ethics into my career an area of concern for me?

4. What type of work do I want to do—for example: lab work, sales, production-oriented work, education and training, policy work, planning, or something else entirely?

5. What type of work environment do I want to be a part of? Do I want to work primarily in an office or out in the field?

6. Do I prefer to work on long-term projects or do I prefer a more loosely defined job with mainly short-term projects and diverse day-to-day tasks?

7. Am I willing to work in a place where I will have to deal with bureaucracy and a paper chase?

8. Do I prefer working in an urban or rural place?

9. Am I willing to travel?

10. Am I willing to relocate to areas where jobs are available?

11. Can I live on a modest but comfortable income, or do I want to quickly climb the financial ladder?

12. Am I willing to go through years of post-secondary education and/or technical training?

It would be a good idea to write down your answers to these questions to get you thinking about what kind of career and life-style you really desire. In order to be truly satisfied with any career, it is essential to first understand yourself and what motivates and excites you. You will be spending a substantial portion of your waking hours at your job, therefore a good match between your personality characteristics and job requirements will make your career a much more satisfying one.

Types of Environmental Opportunities

Environmental Studies

In response to the increased demand for environmental professionals, many colleges and universities across the nation are designing curriculums and degree programs with an environmental focus. There are all types of colleges offering courses and degrees for the career-minded environmentalist. In Chapter 2, traditional majors like forestry and wildlife management will be discussed as well as newer, environmentally focused majors like environmental engineering, public policy, environmental health, environmental law, and interdisciplinary studies.

We will outline some of the major undergraduate and graduate programs that already exist and offer information on new programs and the ways to research degree programs that fit your needs. Courses of environmental study from small colleges like College of the Atlantic, a nontraditional liberal arts college in Bar Harbor, Maine, and larger universities like the University of California at Santa Barbara, as well as technical programs at community colleges, will be discussed in order to give readers a sampling of the many different types of educational experiences. In addition, we will discuss strategies in choosing a college, alternative study programs, and the various financial resources available to the college-bound or returning student.

Jobs in a Greening Corporate America

Environmental employment in the private sector is becoming more and more of a reality as the new environmental consciousness melds with our daily lives. Although many environmentalists feel at odds with corporate America, we have made the decision to place this chapter near the beginning of the book for two reasons. First, the corporate sector of environmental careers has a unique opportunity inherent within it; the private sector has the greatest human and financial resources to impact positive environmental change. Since private companies are often the culprits in environmental disasters, the placement of well-educated environmentalists at the root of the problem and thus, the building of a unified front in the war against destructive human practices gives society the greatest weapon in the fight against environmental degradation. Second, this is the job sector in which, by far, the greatest number of environmental jobs exist. Moreover, given the strong public sentiment about the environment, the corporate sector is under intense pressure to expand the number of environmentally oriented jobs.

Environmental professionals work at the corporate level in areas ranging from policy making, to public relations, to the overseeing of production, and the overall management of environmental programs, among others. Very often, these people are responsible for research into areas of environmental concern, such as implementing and managing work teams and disseminating information to the public, and for proposing and fighting for policy changes. Environmental communications specialists and environmental managers are just two types of careers we will be examining in depth in Chapter 3.

Jobs in Government

In Chapter 4 we will be examining environmental jobs at all levels of government, and we will focus on some of the major federal environmental branches like the Environmental Protection Agency, the National Park Service, and the Forest Service. An environmental career in the government requires a combination of skills and experiences not the least of which are good communication skills, flexible thinking, and patience.

Numerous environmental careers exist at the federal level, in dozens of departments, agencies, commissions, and bureaus. The federal government, in response to public environmental concern, is actively deploying teams that are responsible for developing broad regulatory guidelines, overseeing research, providing technical assistance to state and local governments, and overseeing the enforcement of environmental regulations.

Depending on your level of education and relevant work experience, salaries range from slightly below to competitive with starting salaries in the private sector. In addition, benefits such as vacation and sick time, health insurance, and retirement plans are typically superior to those found in the private sector.

Federal environmental regulations are passed on to state governments, which are responsible for their implementation and enforcement. State governments typically go beyond federal regulations and take initiatives in matters not covered in federal statutes, such as land use, growth planning, and groundwater protection. Numerous state offices and research laboratories carry out specific programs, distribute funds to municipalities, and conduct statewide planning projects. Salaries are typically less than at the federal level and can vary widely from state to state. Job benefits are also quite attractive for state employees and are typically superior to those found in the private sector.

Because of the increased concern about the state of our environment, the initiative to deal with environmental issues is also increasing at the local level. There are a vast number of jobs at the county and municipal levels. Environmentalists working at the community level usually have very hands-on jobs, ranging from inspecting wastewater treatment systems, developing recycling programs, maintaining public land and water use areas, and mediating between developers and residents. In general, the pay and benefits are lowest for this level of government and the atmosphere is more politically charged. Because of the high turnover rate and hands-on emphasis, this is an excellent place for aspiring environmental professionals to start a career. Entry-level professionals are quickly given substantial responsibility and learn the building-block environmental skills that are useful throughout their careers.

Environmental Entrepreneurs

The individual with an entrepreneurial spirit has an opportunity to produce, package, or sell items that make the statement that there are many products that are good for the environment. The eco-entrepreneur, from the environmental artist who makes art from second-hand objects to the environmental vendor who packages or sells goods that are completely recycled or environmentally safe, has a unique opportunity because the public is increasingly concerned with using products that are considered environmentally friendly. In addition, there are an infinite number of possibilities in areas such as recycling, inventing, and providing people with alternatives to chemically laden or wasteful products. There are some wonderful success stories like The Body Shop, Tom's of Maine, and Ben and Jerry's that have set an example to the rest of the world on how to run a successful environmental business.

We will profile a few of these eco-entrepreneurs and give information and tips on ways to be a successful businessperson. Some resources that may be indispensable to the eco-entrepreneur will also be presented.

Environmental Organizations

One of the best ways to decide if an environmental career is the right one for you is to volunteer at an environmental organization. Since there are relatively few staff positions compared with volunteer positions at environmental organizations, we think it prudent for the interested environmentalist to spend some time volunteering or interning with one of these groups. This experience will give you the opportunity to get a feel for the working conditions and atmosphere at nonprofit organizations, and the environmental field in general.

In Chapter 6 we will list and briefly describe some of the larger national environmental organizations and provide information on the ways to get involved with state and local environmental issues. Information on how to specifically pursue volunteering and staff opportunities will also be discussed.

The work done by nonprofit environmental organizations is exciting and the atmosphere is dynamic. While this is the smallest job sector discussed in this book, in the past ten years nonprofit environmental organizations have experienced tremendous growth, and the number of internships, volunteer opportunities, and staff positions has steadily grown. This is also an important career step because it provides you with the opportunity to meet and interact with environmental professionals who could further your own environmental career.

Jobs in the Media

The interpretation and dissemination of information has always been a crucial aspect of the democracy in which we live. When important issues are being discussed and decisions are made in the corporate board room, at an international environmental conference, or on Capitol Hill, it is imperative that citizens have access to information that may have a direct impact on their lives. Choosing a career in the media requires that the individual have a strong ethical perspective and that she or he responsibly investigate and present factual, unbiased information to the general public.

More and more, we are seeing environmental reporters being featured on news shows and in newspapers and magazines. Furthermore, in the last few years, environmental news and information segments have appeared on many local television news programs and on such news and entertainment stations as MTV, CNN, and the Arts and Entertainment Network. Numerous new books and magazines exist that focus exclusively on the environment. All types of personnel like newscasters, writers, editors, camera people, and supporting research staff are needed.

These professions are fast-paced, exciting, and personally rewarding with new opportunities arising daily. Whether you are the environmental reporter for the *New York Times* or a local TV station, your impact is potentially enormous. There is a large and growing audience that wants to be educated on environmental issues. In addition, beyond the confines of your desk, the opportunity for free-lancing your writing skills is also growing. With initiative and resourcefulness, a media environmentalist can earn a handsome living.

Environmental Education

One of the most vital components in making your environmental career a successful one is choosing the correct course of study. This applies not only to recent high school graduates who are considering a college education but also to individuals who have turned their professional attention toward an environmental career. In many organizations today, environmental literacy is becoming a managerial requisite. Many young college graduates majoring in environmental studies are being hired, and an increasing number of professionals are returning to school to learn more about environmental issues.

The good news is that environmental employment opportunities are at an all-time high. Ten years ago there were only a handful of environmental businesses, while today the Department of Commerce estimates that there are between 60,000 and 70,000 companies, organizations, and governmental agencies with an environmental bent. It used to be that anyone with a general interest and some perseverance could break into the environmental field. Today, with the slowing in job growth and a huge increase of interest in environmental careers, the job horizon is getting much more competitive. A recent study at the

University of California-Los Angeles found that 83 percent of all college freshmen "want to make a positive contribution to the environment," and 21 percent "want to work on environmental issues." Couple this with the fact that each year, thousands of people with solid professional experience are switching to environmentally oriented careers, and the job picture zooms into focus and boldly reads: Environmentalism is becoming serious business and only those with a serious interest and sharp skills need apply!

These words of caution are not meant to discourage, because the environmental field is expected to keep expanding well into the next century and needs thousands of qualified people each year to fill its ranks. It doesn't matter if your education, interests, or work background are in the humanities or the basic sciences. There are English majors heading hazardous waste clean-up firms, political scientists setting state and federal recycling policies, engineers developing and marketing high efficiency industrial equipment, biologists leading international conferences on world resources conservation, and business majors working at nonprofit environmental organizations. What made the difference for these people is that they knew how to make the best of their education and skills, and they are now working to save the earth.

Recent Trends in Environmental Education

There are some recent trends in the environmental job sector that deserve particular attention. First, a college degree is becoming a standard criterion for employment. This is a solid requirement for young, entry-level applicants and a definite edge for those with established job experience. A four-year bachelor's degree is becoming the norm for most professional job candidates while a two-year associate's degree or certificate of training from a community college or vocational institute is the prerequisite for most technical positions. It is also not unusual for employers to hire a large number of candidates with some form of graduate training. Increasingly, private companies and government agencies are requiring, as a condition of employment, that their employees take continuing environmental education courses. These classes are partially or fully paid for by an employer and often lead to a technical certificate or even a graduate degree. The reason for these rigorous educational standards is quite simple; the complexity of environmental problems and level of technological and organizational sophistication necessary to solve our most pressing environmental problems make those with a postsecondary education, and a willingness to keep on learning, the most attractive job candidates.

Second, a basic understanding of the physical sciences for all career-minded environmentalists is becoming essential. This may not be a problem for some people, but there are others for whom just the utterance of the words math and science sends a chill straight down their spines. Remember, there is a big difference between an understanding of scientific issues and the pursuit of a degree in the "hard" sciences. By definition, ecology is the study of the relation of living organisms to their environment or the study of ecosystems. In order to grasp the complexity of environmental problems, an understanding of the principles of this relationship is essential. There is ample room in this field for both generalists who may be involved in policy decisions, fundraising, or public education, and specialists who design specific systems or explore particular scientific ideas, problems, and solutions. Each person should have a solid understanding of environmental issues and be able to verbalize the way in which humans have caused harm to the environment and the way that we can work to solve these problems.

There is presently a career advantage for those whose education leans toward the physical sciences. A majority of environmental job descriptions stress some type of specific scientific training. This heavily biased emphasis on training in the sciences does, however, appear to be giving way to a more integrated approach, incorporating both the physical and social sciences. A combination of scientific and administrative skills is beginning to define the ideal environmental career, and those best able to synthesize these skills will be highly sought after. One college brochure describing its environmental science program states this case.

> Society needs specialist biologists, geologists, and geographers, and will continue to do so in the foreseeable future. Society also needs generalist environmental scientists who can coordinate and interpret data emanating from widely differing disciplines. This is vital if man [sic] is to take the necessary steps to halt the current over-exploitation of the Earth, and is to attempt to meet the demands of a human population far in excess of ecological balance . . . it is desirable that a higher proportion of our future politicians and administrators should make environmental science, rather than an arts subject, their basic education.

Finally, there is increasing attention being paid to prospective employees with good reasoning and communication skills. These are the liberal arts tools like analytical thinking, writing, speaking, and cooperative abilities. While these are the strengths of most students interested in the humanities, they have historically been a weakness for those trained in the physical sciences. Increasingly, employers are looking for people who have a substantive understanding of technical issues, who can come up with creative solutions, and most importantly, who can relate their thinking to others. In the CEIP Fund's *Complete Guide to Environmental Careers* the vice-president of a large timber firm pointedly addressed this issue by saying, "What separates the forest managers from the technicians is not their knowledge of forestry but their liberal arts skills: They can work with people, they can communicate, and they can see and solve problems." There is a definite trend toward seeking professionals with more well-rounded education. Those able to master these skills will surely be the environmental leaders of the future.

Sources of Information

Careers for Nature Lovers and Other Outdoor Types (1992). By Louise Miller and published by VGM Career Horizons, National Textbook Co. Includes information on careers in the biological sciences, agricultural sciences, land planning, forestry and conservation science, geology, and pollution control and waste management.

The Complete Guide to Environmental Careers (1989). By the CEIP Fund and published by Island Press. An excellent resource for scientific and technical careers. Included is a thorough discussion of both formal and informal educational information. Book due to be updated in 1993.

Exploring Environmental Careers (1985). By Stanley Jay Sharpiro and published by the Rosen Publishing Group. Contains detailed information on scientific and engineering, technical, and skilled craft environmental careers.

The Basics: A Good High School Education

You can start to prepare for your environmental career long before choosing a college major or trying to convince an interviewer, while squirming in your chair and picking imaginary lint from your clothing, that you are the perfect job candidate. This doesn't mean that you must know your exact career ambitions by the time you enter or leave high school. It does imply that you should take your high school education seriously and use it to your advantage later in life. Many of the basics taught in junior high and high school will make your college experience easier and more rewarding. Those with the best college experiences, in terms of not only grades but also the development of a healthy self-esteem and intellectual curiosity, will surely stand out among the career seeking crowd.

First, a solid grasp of math and English are really the golden keys to your educational and career success. Mathematics is the foundation of all the sciences and is an indispensable skill for any career-minded environmentalist. You should take a full four years of high school math with two years of algebra, one year of geometry and trigonometry, and one year of calculus. The communication arts—reading, writing, and speech—are equally important. No matter what you later choose as a field of study or occupation, it cannot be overstressed that English skills are an essential part of the formula for professional success. In addition to the required English classes, you should take literature, composition, and public speaking courses.

All young, environmentally minded students are advised to take high school courses that offer a wide variety of environmental/scientific information. Classes in biology, the earth sciences, physics, and chemistry are a good start. Along with later helping you understand complex issues like hazardous waste problems, global warming, and ozone depletion, these courses will prepare you for similar but much more challenging college classes. Finally, talk to your guidance counselor and let him or her know what your interests are, so he or she can help you design a plan of study.

Choosing the Right College

Unlike other fields of study such as English, engineering, sociology, or education, environmental studies does not fit neatly or exclusively into one academic department. Only very recently have colleges created environmental programs within academic departments or interdisciplinary programs that combine the teachings of different fields. This is mainly a reaction to growing student interest in environmental studies.

In the past few years, educators have been seriously interested in creating more of these types of programs. Unfortunately, ongoing budget problems have forced many schools to scale back existing programs and defer or cancel new programs altogether. While there is definitely a need for a greater variety of environmental programs at all types of higher learning institutions, it appears the demand will outpace the supply for at least the next few years. Fortunately, there are already a large number of programs throughout the country from which to choose.

Some Important Issues

It would be nice if everyone could find the environmental program and school that fit their needs exactly, but unfortunately this is not always the case. No matter if you are a first time or returning student, you will be faced with making important and often life altering decisions for the sake of your education. The pursuit of higher education requires you to commit a sizeable amount of time, energy, and money in return for a degree that will open new, exciting career doors. Most of us would love to attend the best school, remain near our friends and family, make outstanding grades, and have ample time to study and socialize. Not very often, though, are all of these wishes realistic or even possible. Before researching and collecting information on schools, it would help to consider the following three questions to better understand what you are looking for in a school and an environmental program.

1. What type of environmental training do you really want or need? This is probably your most important consideration in choosing the right school and/or educational track. Do you want a full four-year degree at a university where you will be required to take a wide range of courses in the humanities and "hard" sciences and gain a broad understanding of environmental issues, or do you want a more specific technical or scientific education in two years or less?

2. How much do you want to spend on your education? In general, the least expensive schools are community colleges and technical institutions. Public four-year colleges and universities occupy the middle price range, and private schools tend to be the most costly. Also, it is usually much less

expensive to attend a public college or technical school in your own state. Students from other states are typically charged twice the amount or more than state residents.

3. Would you be comfortable at a large university where classes are large and the atmosphere is somewhat impersonal yet exciting, or would you feel better attending a smaller school where classes are small and the atmosphere is highly personal?

Answering these questions will help you choose the type of learning environment that is right for you and will aid you in choosing the type of school in which you feel the most comfortable.

Two-Year Schools: A Technical Environmental Education

Two-year colleges are a convenient, affordable, and a rewarding way to continue your education and are often a direct conduit to an environmental job. The term two-year school is an umbrella term for junior and community colleges, private occupational schools, area vocational schools, adult education centers, and correspondence schools. These schools enroll more than six million students per year or 40 percent of all college students. The reason for this high enrollment rate is quite simple; our technological society demands a large number of highly skilled workers to keep it functioning. For example, for every one professional job in the environmental sciences, there are three technicians employed.

These schools differ from four-year liberal arts colleges primarily in that they are oriented toward teaching students specific skills that are directly related to employment opportunities. Two-year schools are also a good choice for those who don't feel that the academic program at a four-year college is quite right for them. For example, high school students who are absorbed in one subject like science, mechanics, or the arts, returning students who are seeking further training or a certification for a specific job, or individuals looking to be retrained in order to change careers are all well-suited for this type of education.

Tuition at two-year public schools is one-half or less than tuition at public four-year colleges and one-eighth of the cost of private colleges. Furthermore, many two-year schools have open-door admission policies that make them accessible to everyone. Also, these colleges are community based, which makes them convenient for students who need to maintain a close link with family or employers. Unlike four-year colleges where you may be forced to travel far to find a specific program, two-year schools are much more numerous, and the chances of finding just the right program close to home are quite good. Most community colleges also offer liberal arts transfer curriculums that prepare students with the first two years of courses, after which they can transfer to a four-year school and receive a baccalaureate degree.

In the environmental field, workers with many forms of technical training are in high demand. In the areas of solid waste management, hazardous waste management, air and water quality, and land and water conservation, just to name a few, there is presently a shortage of qualified technicians. In addition, the work generated by the Superfund hazardous and toxic waste clean-up program and military munitions plants alone will employ thousands of new workers well into the next century.

Sources of Information

Peterson's Guide to Two-Year Colleges (annual). Published by Peterson's Guides. Lists over thirty majors that are related to the environment and includes a two-page description of most schools.

Lovejoy's Career and Vocational School Guide (1991). Published by Prentice-Hall.

The Blue Book of Occupational Education (1991). Published by MacMillan Publishing Company. Lists schools both by state and type of occupational education program.

National Council of Trade and Technical Schools, 2021 L St. NW, Washington, DC 20036. A good place to write for additional information on schools or specific programs.

Environmental Learning: The Traditional College Major

Environmental courses are sewn into a wide range of educational offerings. Many of these programs offer a high quality education, and above all, the benefit of a college program that is established, respected, and thus marketable. Typically, students receive either a Bachelor of Arts or a Bachelor of Science degree and choose a major course of study in one academic department.

Peterson's Guide to Four Year Colleges lists over thirty different academic majors with an environmental focus. According to *Barron's Profiles on American Colleges*, there are a little over 230 four-year schools that offer an undergraduate major in environmental studies. The bulk of these programs are in environmental science and are rooted in the basic sciences, but many have an interdisciplinary focus. In addition, there are 28 degree programs specifically in environmental engineering. In the social sciences, environmental degree offerings are a little more sparse. There are 15 colleges offering a degree program in environmental design and three that confer environmental education majors. There are also a number of programs in public policy, geography, and sociology.

Some Popular Environmental Majors

In the agricultural sciences, which focus on the study of plants and animals, environmentally focused majors include fish and game management, forestry, animal and food science, natural resource management, range management, and soil science.

The biological sciences, which concentrate on the study of all living organisms—from humans to microbes, their life processes, and evolutionary development—offer environmentally focused programs in biochemistry, microbiology, molecular biology, ecology, zoology, marine biology, and botany.

In the earth sciences, where the processes that create, sustain, and change the earth are the focus, there are such degree offerings as chemistry, geology, geography, physics, meteorology, and oceanography.

In engineering, the focus is on the application of mechanical principles to practical situations by the use of tools and machines. Environmental engineering is defined by the American Academy of Environmental Engineers as:

> The management and optimum use of air, water, and land resources, and the provision of facilities and the control of conditions for living, working, and recreation that will contribute positively to human health, comfort, and well-being.

Working closely with engineers are surveyors, architects, and landscapers who help design and implement environmental projects.

There are also a host of liberal arts departments that have integrated environmental issues into their curriculums. In sociology, political science, psychology, public policy and administration, education, and business, the focus is on the behavioral aspects of human impact on the environment. Individuals with social science backgrounds often manage environmental projects and conduct policy research.

There are also concentrations in the health sciences, which provide evidence of the medical effects resulting from environmental modification. Some of the medical concentrations include public health specialists, occupational hygienists, emergency/disaster scientists, and toxicologists.

Sources of Information

The following is a list of four-year college directories that are available at libraries, high schools, and in the reference section of many bookstores.

Peterson's Guide to Four Year Colleges (annual). Published by Peterson's Guides, Princeton, NJ.

Barron's Profile of American Colleges (annual). Published by Barron's Educational Service Inc.
Lovejoy's College Guide (annual). Published by Prentice Hall.
Index of Majors (annual). Published by the College Entrance Examination Board, New York.

Interdisciplinary College Majors

There has been a move in many schools to bring together faculty from different departments to form interdisciplinary environmental programs. The purpose of these programs is to integrate the teachings of various disciplines and present the student with a more complete understanding of environmental issues. Remember, the definition of ecology, with its emphasis on the interrelation of humans to their environment, suggests that the social and physical sciences be pulled together and studied simultaneously. In this vein, for example, it makes little sense to study solutions to air pollution exclusively from an engineering standpoint. To do so would disregard the importance of the economic, social, or political factors, all of which influence the outcome of any program designed to abate air pollution. It may be technically feasible to place scrubber devices on all smokestacks and drastically curtail the amount of toxins released into the air, but it may be difficult, if not impossible, to get companies to actually install them without government regulations. What appears as a rather simple task from an engineering standpoint is actually a challenging objective when considered in the economic and political realm.

The intent of interdisciplinary programs is to overcome the shortfalls of unidimensional classroom learning. A course that considers the subject of air pollution control would have economists, sociologists, and political scientists, as well as engineers, explain their various approaches to the problem and would give the student a more realistic conception of how environmental problems are approached and solved. This combination of technical, social, political, and economic teachings is becoming the core of many environmental programs.

Some large universities that have developed solid interdisciplinary environmental curriculums are **The Pennsylvania State University,** State College, PA 16801, which, through the cooperation of the departments of meteorology, geology, and geography, operates the Earth Systems Science Center. At the **State University of New York** at Syracuse, NY 13210, the College of Forestry and Environmental Science has built a good reputation. The **University of North Carolina at Chapel Hill,** Chapel Hill, NC 27512, has built its department of environmental science and engineering into one of the finest programs in the country. **Indiana University** at Bloomington, IN 47401, offers environmental studies in both the College of Liberal Arts and the School of Public and Environmental Affairs, and **Illinois State University** at Normal, IL 61761, has an environmental health program that integrates medical and public policy issues. Also, the **University of California at Santa Barbara,** Santa Barbara, CA 93106, has a fine environmental studies curriculum in its School of Liberal Arts.

Small and Innovative Environmental Schools

Most of these schools offer an alternative education experience, where students are encouraged to be independent and creative thinkers and are individually guided to pursue the environmental ideas and issues that interest them most. Many of these environmental colleges were established in the 1960s and 1970s, during the first wave of environmentalism, and their continued success exemplifies a deep commitment for environmental values shared by succeeding generations.

Students are motivated by curiosity rather than grades, and a good number of these schools have done away with the rigid grading system in favor of faculty and peer review of the students' progress. Learning is demonstrated through small group interactions, individual projects, field work, and internships. The major advantage of these small and focused schools for the student is the faculty's deep commitment to teaching. The following is only a sampling of the many educational institutions that have respected programs and motivated and dedicated students and faculty.

Prescott College is nestled in the wooded mountains of central Arizona in a small, close-knit community that provides students with a healthy atmosphere and a strong sense of "commitment to responsible participation in the natural environment and human community" (from their general catalog). This is a small school with a little over 700 students living in and around the campus. Prescott College offers Bachelor of Arts degrees in environmental studies, human development and education, outdoor education and leadership, cultural and regional studies, and humanities. During the first month of residency, all new students participate in a three-week wilderness expedition designed to teach environmental ethics and reverence for the planet. There is also an Adult Degree Program through which adult students can earn a bachelor's or master's degree in liberal arts fields including education, counseling, management, psychology, and human services. Students complete their degrees in their own communities and the only residency requirement is two weekends at the Prescott campus.

For more information about the college, write or call Prescott College, 220 Grove Avenue, Prescott, Arizona 86301, phone (602) 778-2090.

College of the Atlantic is a four-year, independent college located on Mount Desert Island in Bar Harbor, Maine. The campus consists of 26 shore-front acres adjacent to Acadia National Park and overlooking Frenchman Bay. The school offers both a Bachelor of Arts and a Master of Arts degree in ecology. The teaching of the interconnection of humans with their physical surroundings is a central mission of the college. The curriculum is split into three areas of concentration: environmental science, arts and design, and human studies. The college has a strong concentration of classes in marine biology, environmental design, environmental media, and education. Undergraduate requirements include at least two courses in each area of concentration, a human ecology essay, an internship, a senior project, and community service. The master's degree program is a more intensive extension of the undergraduate program, and many students who enter the program continue with their undergraduate focus.

There is a cooperative agreement between the college and Acadia National Park, where students conduct much of their field work. The college also maintains the Island Research Center where, "students monitor populations of endangered or threatened bird species, develop censoring techniques for bird populations, and observe the impact of changes in island vegetation on animal species" (excerpted from the college bulletin).

For more information write or call College of the Atlantic, Bar Harbor, Maine 04609, phone (207) 288-5015 or (800) 528-0025, or fax (207) 288-2328.

Evergreen State College is a four-year public school tucked away in the rolling and forested mountains of Olympia, Washington. This is one of the very few state schools that has both an alternative curriculum and an environmental focus. Evergreen offers both a Bachelor of Arts and Bachelor of Science degree, as well as a master's degree in environmental studies. The environmental studies program stresses the interaction of human societies and nature, and the goals of this program are:

To understand the nature, development, and interactions of terrestrial and marine ecosystems and human societies;

To learn the richness and the limits of the environment and social resources available to sustain both human environments and natural systems;

To study the cultural values and philosophies that shape environmental behavior; and

Through applied work, to develop the skills necessary to handle our resources wisely.

The environmental studies program also maintains a close working relationship with the political economy and social change and science, technology, and health programs. The master's degree program combines public policy and environmental science so that graduates can use a combination of management and technical skills.

A student's academic pathway is at first structured, with a number of core program requirements, but later becomes more independent and specialized. Instead of grades, faculty members write a narrative evaluation of each student's work and in turn, students prepare a self-evaluation and also one of the instructor. Thus, Evergreen stresses learning through an open and honest two-way communication system between faculty and students.

Further information can be obtained by writing or calling the administrative office at Evergreen State College, Olympia, Washington 98505, phone (206) 866-6000.

Other notable small environmentally focused alternative colleges include **Hampshire College** in Amherst, Massachusetts 01002; **Williams College** in Williamstown, Massachusetts 01267; **Colorado Mountain College** in Leadville, Colorado 80461; and the **Huxley College of Environmental Studies,** a unit of Western Washington University in Bellingham, Washington 98225.

Additional Sources of Information

The Right College (1991). Published by Arco Press. A reference guide that lists 55 interdisciplinary environmental study programs.

Opportunities in Environmental Careers (1992). By Odom Fanning and published by VGM Career Horizons, the National Textbook Co. This book has an in-depth chapter on educational opportunities focusing on alternative schools.

Environmental Study Alternatives: Making Your Own Major

While there are indeed many more degree programs in environmental studies today, as opposed to ten years ago, finding a school that matches both your interests and needs may still be a big challenge. A student interested in an environmental engineering program may find that there are none in her or his state, or someone interested in both ecology and public policy may have trouble finding a school that has such a joint program. In addition, students are often faced with making difficult life-style compromises. Colleges that have an established environmental studies program may be far from home or too expensive and thus not a practical choice. When faced with this dilemma, some may decide to compromise and choose a program that isn't quite what they wanted.

A good alternative, which requires a flair for creativity, is to design your own special environmental major. In this case, it is up to you to seek out schools in your area that have the relevant environmental courses, to create a class schedule for each term, and meet all the course requirements of the school. Many schools permit this type of flexible scheduling, especially when the student has specific interests and goals.

There are usually a number of faculty members at any school pursuing environmental issues. It would be wise to find out who these faculty members are and contact them to see if you have similar interests. Also, make sure that they are interested in what *you* want to do and most importantly, have the time to help you. Contact the dean's office of student affairs to see if this type of arrangement is feasible.

Graduate Schools

Graduate training is far and away the most popular trend for workers in the business world today. For example, how many people do you know who have a Master's of Business Administration (MBA) degree or are planning on attending graduate school? Just a few years ago, holders of advanced degrees were the exception—now they are becoming the norm.

The environmental field is ripe for professionals with advanced educational training. This field has experienced tremendous growth in the past ten years and requires individuals with advanced scientific training and modern management skills. A recent survey of environmental clean-up firms found that an increasing number of their new professional employees have advanced degrees in fields like biology, engineering, chemistry, and business administration. In addition, nonprofit environmental organizations are hiring a large number of both scientists and policy analysts with graduate degrees.

Graduate school can be a wonderful learning experience and very helpful for your career, but you should think long and hard about your motivations for wanting a graduate degree and the type of training you hope to receive. Unlike most undergraduate programs, which have their core subject matter and offer a general learning experience, graduate programs are intensive and specific.

If you have just finished your undergraduate education and are torn between going directly to graduate school or first getting some work experience, most college faculty find that returning students are generally more focused and successful because their studies are tied directly to their careers. This doesn't mean that you should not go directly from undergraduate to graduate school, but today many students enter graduate school as if it were an extension of their undergraduate training, with little if any idea of what they would like to do after finishing their degree. Far and away, the most complete and detailed information source on graduate programs is *Peterson's Annual Guide to Graduate Study*, which comes in five volumes and gives a detailed description of schools, programs, departments and faculty, and their present academic focus.

Vol. 1- Graduate and Professional Programs, Overview

Vol. 2- Humanities and Social Sciences

Vol. 3- Biological, Agricultural, and Health

Vol. 4- Physical Sciences and Mathematics

Vol. 5- Engineering and Applied Sciences

A Cautionary Note on Graduate Programs

While the need for integrated environmental learning is well recognized in the "real world," there is still quite a bit of resistance to forming interdisciplinary programs at our institutes of higher learning, especially in graduate programs. The reason for this is unfortunately quite simple: many professors consider interdisciplinary studies as academically inferior to the work done in their own disciplines. Call it a form of academic xenophobia, but theory has it that incorporating other subjects into a discipline somehow "waters it down" and makes the work less respectable. This stigma also carries over to students. Their broad academic interests are seen as academically diluted, and they are often viewed as being less bright than their more specialized counterparts who study a single subject. This is not the norm in all graduate programs but it is something to be aware of. When researching graduate programs, make it clear that you seek additional training for professional reasons, that you intend to return to the real world, and that you are not interested in a career in academia, where the stigma is strongest.

Law Schools

In recent years the battle against environmental degradation has increasingly been taking place in the courts of law. In the 1980s, environmentalists, frustrated with political foot-dragging in the enforcement and reauthorization of environmental regulations, looked to the courts for protection. The result of all this activity was a huge increase in litigation and the need for a larger corps of lawyers with environmental training. There are still only a handful of law schools that offer training in environmental law.

Northwestern Law School is a fully accredited law school located in Portland, Oregon. Northwestern has an Environmental and Natural Resource Program in which more than one-third of the school's students participate. To earn a certificate and claim a concentration in environmental and natural law, students are required to take, in addition to their required classes, several courses in environmental law and write two publishable papers. Also, *Environmental Law*, a journal devoted to environmental law issues, is published at the school. Many of Northwestern's graduates go on to work for national environmental groups like the National Resource Defense Council, for governmental agencies, and for socially conscious private law practices.

For further information write or call the Northwestern Law School, Department of Admissions, 10015 SW Terwilliger Boulevard, Portland, Oregon 97219, phone (503) 244-1181.

The University of Colorado Law School, located at the foot of the Rocky Mountains in Boulder, Colorado, established the Natural Resource Law Center in 1981 because of the steady rise in legislation and regulations regarding natural resources. The center hosts educational programs for scholars, government officials, public interest groups, and members of industry as well as sponsoring distinguished visitors and supporting research and publications pertaining to natural resources. While the center sponsors no formal education program in the law school, students gain constant access and have the opportunity to foster professional relationships with practicing lawyers and scholars who deal with natural resource issues.

For further information write or call the Natural Resources Law Center, University of Colorado School of Law, Campus Box 401, Boulder, Colorado 80309–0401, phone (303) 492-1286.

Sources of Information

Barron's Guide to Law Schools (annual). Published by Barron's Educational Service, Inc., New York. Contains information on environmental law programs and the way to prepare for law school as an undergraduate.

Association of American Law Schools, 1201 Connecticut Ave. NW, Washington, DC 20036. Write for a listing of accredited law schools with environmental programs.

World Environmental Directory (annual). Published by Business Publishers Inc. Contains a listing of attorneys with environmental interests.

Loans and Scholarships

The cost of a college education has more than doubled in the last ten years and is expected to rise at an even faster rate in the coming decade. It was not long ago that only students interested in private schools or specialized programs like law and medicine had to consider cost as a deciding factor of college attendance, because the vast network of public schools offered a high quality education at an affordable rate. This simply is no longer the case. According to the National College Board, a typical four-year public college education including tuition, room, and board averages $5,656 per year and a whopping $12,874 at private colleges (tuition per year alone averages $2,253 at public schools and $8,879 at private institutions).

Student Loans

Loans and scholarships are the two main sources of financial aid available to all students. Student loans are relatively straightforward and easy to secure. Almost every school has a financial aid office or representative where you can get information on the amount and terms of loans, as well as help with filling out the required forms. The majority of students seeking money for school use these loans as their main source of financial aid. The largest loan programs in the country are the Stafford Loan Program (formerly the Guaranteed Student Loan Program) and the Perkin's Loan Program (formerly the National Direct Student Loan Program). Student loan programs once had the reputation of being a source of easy money because many students simply did not repay their loans. The government and banks are now much more strict with their lending policies. The IRS is now freezing offenders' bank accounts and forcing them to repay their defaulted loans.

Scholarships

Scholarships are money *given,* not lent to students. Money is granted to those who, on the basis of need or merit, meet the requirements of the granting body. There are literally thousands of organizations with scholarship funds throughout the United States, and every year thousands of individual scholarships, worth millions of dollars, go unused simply because no one bothered to apply for them. Many students are under the incorrect impression that scholarships are only for the poor and academically or physically gifted. You can receive a scholarship because of particular interests like writing, reading, or hobbies, the geographic area in which you were raised, special skills, language abilities, mechanical inclinations, and so on.

For just about any human ability, there is a scholarship out there just waiting to be awarded. At Juniata College in Pennsylvania, for example, left-handers compete annually for a $20,000 grant left by a southpaw couple. Volunteers for the Student Union Social Committee at the University of Wisconsin who were raised in the New York area are eligible for a $500 scholarship each year. There are perhaps thousands of grants created specifically for the study of environmental problems.

There are no universal standards for scholarships; the amount of financial award depends solely upon the granting organization. In general, most scholarships range from $50 to $1,000 while some even pay full tuition costs. There is no one listing of scholarships or their sources and consequently, the most difficult part of receiving a scholarship is just finding out about it. You can either do the research on your own or use a scholarship search service. These services typically charge $50 to $100 per search and have the advantage of being able to narrow the scholarship search specifically to awards that highlight your personal strengths and specific sources of environmental funding.

Sources of Information

The Scholarship Book (1987). Published by Prentice Hall. This is a complete guide to private-sector scholarships, loans, and grants for undergraduates. There is a section on funding for environmental studies and the chapter "Special Publications" lists career and scholarship guides for individual careers.

The College Blue Book: Scholarships, Fellowships, Grants and Loans (1991). Published by MacMillan & Company. An extensive listing of sources of student financial aid. Arranged by broad fields of interest, including a section on environmental studies.

Scholarships, Fellowships and Loans (1992). Published by the Bellman Publishing Co. An extensive listing of sources of financial aid with a detailed description of each source. The "Vocational Goals Index" helps identify which scholarship to apply for.

Additional Sources of Information

Educational Resource Information Center (ERIC). Sponsored by the U.S. Department of Education, ERIC is a database that provides computer access to information dealing with education. This system includes over 10,000 documents dealing with environmental education. Check at local colleges and high schools for access to ERIC.

Conservation Directory (annual). By the National Wildlife Federation. Included in this potpourri of environmental information is a chapter on colleges and universities with environmental programs.

Opportunities in Environmental Careers (1992). By Odom Fanning and published by VGM Career Horizons, National Textbook Co. Includes a chapter on educational opportunities.

World Environmental Directory (annual). By Business Publishers, Inc. A large reference manual including entries on environmental education programs, databases, and funding sources.

Meeting Environmental Workforce Needs: Education and Training to Assure a Qualified Workforce. Proceedings of the conference of the same name, April 1985. Cosponsored by the EPA, Tennessee Valley Authority, and *Pollution Engineering* magazine. For a copy write: Information Dynamics Inc., 111 Claybrook Drive, Silver Spring, MD 20902.

THREE

The Greening of Corporate America

As our concern for the environment continues to grow and the clean-up of our environment becomes a political issue and a national priority in the 1990s, many new challenges and opportunities have arisen for corporations and individuals that have not traditionally concerned themselves with the environment. This rising concern about the environment and an increased awareness of the environmental impact of corporate actions by consumers, legislators, and executives, coupled with more and stricter environmental regulations, is forcing companies to rethink their manufacturing, production, and management strategies. The challenges lie in integrating environmental policies without losing jobs and drastically reducing profits, and in reshaping the thinking of corporate and industrial America as it exists today. The opportunities can be found in new emerging markets for environmentally conscious products and increased employment in environmental industry.

The integration of an environmental component into jobs such as banking, law, and marketing, which have traditionally been removed from a concern for the environment, has created a world of new opportunities. People can now combine their concern for the earth's natural resources with their interest in business. Across the country, new positions, new divisions, and new industries have been created. An environmentally literate manager enhances decision making and ultimately designs and implements policies that will eventually provide a new framework for management policies and successful businesses. An environmentally literate retailer will respond to the consumers' request for safe products by developing and marketing products that have been produced without harming the earth. Environmentalists will turn to scientists and inventors to develop innovative and safe products.

Environmental awareness is starting to become an integral part of mainstream corporate culture and will continue to do so. The environment must be considered in the daily decisions of employees and managers in a wide variety of areas from purchasing, packaging, research and development, marketing, sales, and training, to public relations. The division that has existed between the "environmentalists" and the "polluters" since the beginning of the industrial revolution has begun to close, and numerous opportunities are available for those who are willing to close the gap even further.

The new corporate employee is a "green-collar" worker, one who can connect personal values about the environment with professions. People with expertise in the sciences, finance, management, policy making, and communications are increasingly finding jobs that combine their interests and skills. For the job seeker with the right education, experience, and commitment, the financial rewards can be significant.

Opportunities in Green Corporate America

The following list of opportunities for employment in a greener corporate America is designed to give you a general idea of where new positions are being created in response to the demand for greater control over environmental issues. As you will note, the environmental factor is present in almost every aspect of business, and your knowledge of environmental issues coupled with expertise in any given area will make you an ideal candidate to become part of the greening of our nation's businesses.

Public Sector

This group involves federal, state, and local government officials who have the responsibility of regulating industry, allocating resources for clean-up, defending the environment, solving environmental problems, and a number of other specialty professions.

Private Sector

Private interest groups, labor groups, and other nonprofit organizations have grown and have become more corporate in their organizations and management approach. These organizations are hiring analysts and managers to grow and develop their organizations.

Private Industry

The industrialists, the major polluters and consumers of our natural resources, are under increased public scrutiny to try and cut pollution at the source—where the products are made. The primary metal, chemical, petroleum, and paper industries are long-standing industries that all need help in revising and rethinking traditional nonenvironmentally sensitive practices.

Education

Educating the masses about the critical need to clean up, monitor, and safeguard our environment is one of the most important environmental careers, because of the changes in behavior it may cause. Teachers at the elementary and high school levels are helping to create an environmental ethic in a new generation of youth. In addition, consultants and trainers will be needed to educate managers and future business leaders at both the college and corporate levels.

Consulting

As a result of new regulations and consumer demand, many companies are seeking help during transition phases when information is needed to educate management. Technical and environmental management experts work directly with businesses on compliance and proactive management policies.

Law

Environmental law is one of the fastest growing areas of legal study today. Lawyers help draft new legislation for improving the environment and help in holding polluters responsible for their actions. Environmental lawyers rely on the expertise of consultants and engineers to help tighten their litigation.

Science

Scientists from chemists to biologists to anthropologists are continually researching and finding out more about our earth. They assess the type of effects various activities have, and propose solutions for conserving the earth's resources.

Retail

As consumer demand continues to create markets for environmentally sensitive products, new opportunities are being created for new products and packaging. Managers will need to understand the wants and needs of their customers and respond to new legislation that requires more precise terms for labeling and packaging.

Agriculture

New opportunities have developed for those people working in agriculture, due to consumer demand for quality food and content information. New markets have emerged for organic produce and foods, as concern about pesticides continues to grow. A number of new jobs have been created for those with expertise in pest management, organic gardening, retailing organic food, and mail order sales.

Communication

Communicators are needed in both the public and private sectors to explain new policies, highlight the activities of public interest groups, work on technical journals, and report on environmental events and issues. New positions are being created at newspapers across the country for the environmental writer, and the number of environmental magazines and journals continues to grow.

Finance

Professions that have not traditionally been directly concerned with environmental issues such as banking, accounting, and insurance are now finding themselves affected by environmental problems. Banks now have to consider the environmental policies of the companies they lend money to. The growth of environmental funding and investing in service and clean-up industries has created new positions in response to consumer demand for investment opportunities. Insurance companies provide the policies to the firms that are at risk for environmental accidents. As a result, environmentally aware underwriters are now in demand.

Entrepreneurs

The opportunities are endless for the creative, environmentally aware individual or corporation to create new products and new businesses in response to consumer demand. From recycling businesses to environmental cosmetics, the entrepreneur will find a large and receptive market.

CASE STUDY: Canon: "Ecology, Quality, Cost, and Delivery"

In June of 1992, a special "green" advertising supplement to the *New York Times* showcased numerous companies and organizations who have been working toward proving that profits and environmental protection are compatible. One of the most impressive examples illustrating the successful greening of corporate consciousness is Canon, a company with sales of nearly $15 billion in cameras, copiers, and other products.

In 1990, Canon's internal motto of "Quality, Cost and Delivery" was changed to "Ecology, Quality, Cost and Delivery." This shift in the company's overall philosophy placed the preservation of the environment as central priority and has led the implementation of a new set of control standards by which all of Canon's manufacturing and other activities are monitored.

As a result new recycling programs have been implemented, research and development centers have been formed, and money from some of these ventures has been given to environmental organizations. In addition, a new division called "E-business" has been formed to research and develop and produce a new line of clean air products.

Inherent in this change of philosophy is the creation of new positions for a diverse group of people with a wide variety of talents and interests. Environmentally literate scientists, managers, writers, public relations and advertising people are just a sampling of the areas that have become green in order to carry out Canon's new environmental consciousness.

Implications for the Workplace

Canon is an example of one of the world's fastest growing manufacturers that is benefitting financially by managing its company under a new green corporate philosophy. Environmental consciousness is evident in almost every aspect of the organization. Using the Canon model and philosophy of mutually rewarded coexistence and "a serious and international stand on the environmental challenge in the 21st century" as an example, and applying it to the industry as a whole, the following "green" jobs could result.

Educators. In addition to actual staff or consultants who would be hired to educate employees and educators, employment opportunities would also be created in other industries. For example, as with Canon, more funding would be filtered to publishers and networks for the development and production of environmental media. Those organizations would then need to look to the experts in the field as consultants.

Community Relations. Beyond the development of an internal campaign and company philosophy are numerous positions designed for the presentation, identification, monitoring, and relaying of environmental news to the members of the corporation and the general public. This would include public relations managers, writers, coordinators, and a variety of other positions.

Control Standards. The development and implementation of new environmental guidelines and programs would require that managers, planners, engineers, researchers, and governing organizations all work together. Environmental literacy would be a requirement for all.

Research and Development. In addition to the planners and managers who would be involved in establishing new R&D centers, scientists, engineers, and computer technologists would also be needed. If the research was to take place in a foreign country, language and cultural specialists would also be needed both for cultural reasons and educational purposes.

New Businesses. While in the process of reorienting the corporate culture, many new businesses will be started with the environment in mind. Positions will be available throughout these divisions, and they will have as a requirement, environmental literacy.

As you can see, the opportunities are almost endless for new types of employment in a greener corporate America. The direction your career takes depends a lot on your individual interests and area of study. While more and more educational institutions are adding environmental components to their traditional curriculums, there are still many creative ways for you to customize your career by combining your own interests and values.

Finding Employment in Green-Collar America

Before making a decision about where and how you fit into a greener corporate America, you need to evaluate yourself to understand how to best market yourself in areas of environmental employment. The following is a list of things you need to consider before making a commitment to a specific career.

• Values. One of the first issues you will need to examine will be how your environmental values compare to your potential employer's. In addition, because so many areas of the environment need our attention, you will need to decide where you want to focus your efforts. You need to understand and examine how and where your specific skills will be most valuable and useful and how these skills could fit into the overall management of an environmental organization.

• Goals. You need to ask yourself what you want to accomplish by working for an environmentally sensitive organization. Are you willing to work diligently and continuously to better our earth? Are you patient enough to accept each small accomplishment as a very small but important part of the task of cleaning up the destruction of the Industrial Revolution? Are you prepared to fight for what you believe in with large bureaucracies and governments?

• Skills. In order to find a job that has an environmental component, you need to bring two types of skills to the position. You will need to be well-trained in a technical profession such as engineering, marketing, or finance, and also be well-educated in legal, regulatory, scientific, and political issues as they relate to the environment. To do this, you must study, read, volunteer, and educate yourself in the environmental issues and policies that exist today and those that are being considered for the future.

• Research and networking. As you begin your job search you will need to know who is taking the lead in areas that interest you. As a citizen you have access to corporations' annual reports and 10-k IRS filings that disclose environmental matters and expenditures. Networking just means getting to know people who are already working in environmental careers or organizations. In addition to helping you narrow your area of expertise, the people you meet may be the ones who will help you eventually get a job.

Interviewing for an Environmental Job

The reality of the present job market is that for many, the environment is a sensitive issue. At this point in time, an environmental philosophy is relatively new to many corporations and you must acknowledge that at least having a philosophy is an improvement. If you are embarking on an environmental career, you must also be willing to work with the polluters to correct decades of unguarded development. Understanding the complexity of instituting new policies and overall change will get you far in your first interview. Some of the guidelines for a successful interview are as follows.

1. Do your homework. Try to understand generally what the company's past policies have been and how they are evolving into greener policies.

2. Demonstrate that you are up-to-date on environmental issues by knowing what environmental claims the company is making through its advertising. Research these claims and think creatively about how the advertising impacted you. Share your perceptions as a consumer with your interviewer.

3. Acknowledge that no company is perfect and understand and show appreciation for progress that has been made.

4. Do not use the interview to bash the company for past environmental policies.

5. Express your desire to participate in future growth and change.

During the interview you will also want to ask about the company's plans for future growth in environmental areas and about plans for adding new positions, divisions, and strategies for environmental management. In conclusion, what you need to bring to the interview is knowledge, ideas, enthusiasm, and an understanding of the major issues facing corporate America in relation to environmental issues.

Resources to Consult for Planning Your Green Career

The Complete Guide to Environmental Careers (1989). By the CEIP Fund and published by Island Press.

Crossroads: Environmental Priorities for the Future (1990). Edited by Robert Kahn and published by Island Press.

The Green Consumer (1990). By John Elkington, Julia Hailes, and Joel Makower and published by Penguin.

Green is Gold (1991). By Patrick Carson and Julia Moulden and published by HarperBusiness.

Green at Work (1992). By Susan Cohn and published by Island Press.

Environmental Business Journal: Strategic Information for a Changing Society. P.O. Box 371679, San Diego, CA 92137-1769.

Journal of Environmental Education. Heldreff Publications, 4000 Albemarle Street, NW, Washington, DC 20016.

FOUR

Jobs in the Government: Federal, State, and Local Opportunities

The United States has long been recognized as the world leader in environmental protection and preservation. Our nation has the largest body of professionals and resources directed toward environmental concerns and the toughest anti-pollution measures in the world. Americans, it is said, are enamored by nature like no other people on earth. Only baseball surpasses the enjoyment of the great outdoors as a national pastime, and only by a hair! Furthermore, citizens look to the government to keep their natural surroundings pristine and themselves safe from environmental hazards.

Environmental issues, especially those linked with health threats, have in fact become so prevalent in the last few years that the government now claims the environment to be a top national priority. In the last two presidential elections, the candidates have made this issue a major part of their political platforms. In the 1988 presidential contest, George Bush propelled the environment into the political realm when, speaking from a ship in the pollution-clogged Boston Harbor, he proclaimed: "I would be a Republican president in the Teddy Roosevelt tradition. A conservationist. An environmentalist." Four years later, Senator and now Vice-President Albert Gore, a champion of environmental causes, observed, "For all its rhetoric, the Bush administration has been an environmental disaster. The president has ignored the threat of global warming and the depletion of the earth's ozone layer, undermined enforcement of the Clean Air Act, supported drilling in Alaska's precious Arctic National Wildlife Refuge, abandoned his "no net loss" wetlands policy, and opposed efforts to increase recycling." He also observed that Boston Harbor is dirtier now than it was four years ago.

With the election of a president and particularly a vice-president with strong environmental credentials, there are high expectations that the U.S. government will again take the world lead in formulating international environmental policy. Thus, government agencies that deal with environmental issues are expected to be highly active after 12 years of relative obscurity.

A Brief History of U. S. Environmental Policy and Its Effect on the Government Job Landscape

While the environment has only recently resurfaced as a "hot" political issue, federal involvement in environmental protection and preservation has a long and impressive history. The federal government jumped headlong into creating what is now the most extensive national park and forest system in the world when in 1872 Congress made Yellowstone our first national park. In the early 1900s, President Teddy Roosevelt, a great outdoorsman, charted the direction of national environmental policy when he appointed Gifford Pinchot as the secretary of the Department of Agriculture. Under Pinchot's stewardship, the federal government took the lead in developing public lands for both recreational and industrial use. In 1902, the Department of Agriculture formed the National Forest Service and in 1916 the National Park Service was added to the Department of the Interior. With the addition of many national parks and forests in the ensuing years, these departments grew, and the driving philosophy behind American environmentalism was one based solely on resource conservation.

An abrupt broadening of federal environmental policy, fueled mostly by public concern, took place in the 1960s. In 1962, Rachel Carson released her now famous book *Silent Spring*, in which the great dangers of the pesticide DDT were first exposed. This work attuned the public to the harms caused by industrial pollution, and it is credited with igniting the birth of the age of modern environmentalism. What resulted in the following years was a flurry of federal legislative action designed to preserve the environment and, as importantly, to protect public health.

The result of this legislative activity was the broadening of federal involvement in all environmental matters. In order for the federal government to keep up with its new regulatory powers, a large number of new commissions, committees, and offices were formed. In addition, the Environmental Protection Agency (EPA) was created as an independent agency to act as the nation's environmental nerve center.

Since the early 1980s, the federal government has relegated much of the enforcement of environmental laws to the states. Thus, a rippling effect has occurred where the states have also greatly broadened their environmental protection activities and work force. In fact, states like California and New Jersey have enacted environmental protection laws that far exceed regulations already set by the federal government. As the work of state and local environmental agencies grows, there will be a need for a larger and better trained corps of environmental professionals.

There are thousands of job opportunities at the federal, state, and local levels for people with all different types of employment and educational backgrounds. In sum, the outlook for employment opportunities for the career-minded environmentalist at all levels of government is outstanding. A recent study by the Department of Energy projected that by the year 2000, 20 percent of all environmental graduates will work for the federal government and a substantial portion of the rest will find employment at the state and local levels of government.

Pay and Job Benefits for Government Workers

Government employees generally earn lower wages than their counterparts in private industry, but they usually receive better benefits like vacation time, medical plans, flexible work hours, and greater job security. Federal employees receive the highest average pay and are followed, in order, by state, county, and local civil servants. In all levels of government there are strict hiring and salary regulations. While you will rarely be in a position to negotiate salary or job responsibilities, you can be confident that you will be treated fairly and equally with all other applicants. The old anecdote, "It's not what you know but who you know," is much more applicable in the private sector.

In the federal government, the pay schedule or general schedule (GS) is divided into 18 pay steps. Within each GS step, there are 10 pay levels. The actual pay difference between the lowest and highest levels within each GS step differs by about 30 percent. The following is the GS pay schedule as of January 1991.

Step	Lowest Pay	Highest Pay
GS-5	16,973	22,067
GS-6	18,919	24,598
GS-7	21,023	27,332
GS-8	23,284	30,268
GS-9	25,717	33,430
GS-10	28,322	36,818
GS-11	31,116	40,449

Starting level of pay depends on an applicant's level of education and relative professional experience. In general, you can
expect to start at the lowest level of each GS pay step. Those
with a high school or equivalent diploma can expect to receive
GS-5 wages. A bachelor's degree will earn you a GS-7 rating
while a master's degree will secure you a GS-9 rate. Those who
have gone the extra mile and have earned a Ph.D. will be
rewarded with the GS-11 pay rate. In order to jump up one full
GS step, you will first have to go through a complete yearly
performance evaluation. Periodic pay raises within each GS step
are usually a routine matter. State hiring practices and salaries
are structured similarly.

Government Jobs at the Federal Level: How and Where to Apply

There are literally thousands of job openings in the federal
government each year for career-minded environmentalists.
Only about one-third of these jobs are in Washington, DC, so,
contrary to common belief, you do not have to move to Washington when considering a career in the federal government.
There are jobs in federal regional offices in every state, major
city, and in a large number of counties.

Before beginning your job search you must fill out the Standard
Form (SF) 171 application for federal employment. By federal
law, anyone applying for a nonpolitical civil servant position
must submit an SF 171. There are horror stories about individuals who were not able to pursue federal employment because they
could not complete the SF 171! While it is a long and tedious
form—it is five pages and you will need to list every job that you
have held in the past ten years—it by no means lives up to its
horrible reputation.

Since each agency receives thousands of these forms annually, it would be beneficial for you to somehow make your application stand out from the others. Along with each job description or announcement number, in the government's terms, is a quality ranking factor (QRF). This is a short list of qualifications or areas of specific substantive knowledge that the agency is seeking in an applicant. An assistant director in the EPA suggests that applicants write a short paragraph in which they describe their knowledge and/or experience for each QRF point. This should then be used as a cover letter to the SF 171. This method will not only make your application physically stand out from the rest, but it will provide a more specific profile of your qualifications and more importantly, an edge over other equally qualified applicants.

Navigating the Federal Job Maze

Because of the large number of departments, agencies, and offices in the federal government, you could easily get bogged down in a time-consuming and exhausting search of just a fraction of the jobs that match your interests and/or professional strengths. Many job hunters have likened this job search to an unsolvable maze that has only one entrance and no escape. Figure 1 is a breakdown of the environmental responsibilities in the major parts of the executive branch of government. This illustration will give you a clearer idea of the types of environmental problems with which each major executive branch deals.

The next step in the job hunting process is locating the particular agencies in which you may be interested. Fortunately, the Office of the Federal Register publishes *The United States Government Manual,* which is a complete listing of all the agencies, offices, bureaus, and so on, in each department. This manual also lists the private and international agencies that the

Figure 1 Major Executive Branch Agencies with Environmental
Responsibilities

President				

The Executive Office of the President

White House Office	Council on Environmental Quality	Office of Management and Budget
Overall policy Agency coordination	Environmental policy coordination Oversight of the National Environmental Policy Act Environmental quality reporting	Budget Agency coordina- tion and management

Environmental Protection Agency	Dept. of the Interior	Dept. of Agriculture	Dept. of Commerce	Dept. of State
Air & water pollution Pesticides Radiation Solid waste Superfund Toxic substances	Public lands Energy Minerals National parks	Forestry Soil conservation	Oceanic and atmospheric monitoring and research	International environment

Dept. of Justice	Dept. of Defense	Dept. of Energy	Dept. of Transportation	Dept. of Housing and Urban Development
Environmental litigation	Civil works construction Dredge & fill permits Pollution control from defense facilities	Energy policy coordination Petroleum allocation R & D	Mass transit Roads Airplane noise Oil pollution	Housing Urban parks Urban planning

Dept. of Health and Human Services	Dept. of Labor	Nuclear Regulatory Commission	Tennessee Valley Authority
Health	Occupational health	Licensing and regulating nuclear power	Electric power generation

Source: Council on Environmental Quality, *Environmental Quality, Sixteenth Annual Report of the Council on Environmental Quality* (Washington, D.C.: U.S. Government Printing Office, 1987).

government funds or participates with. There is a full description of each agency and an address and telephone number listed for further information. This manual can be found in many local college libraries as well as high school job resource centers.

After researching the agencies that appear to match your interests, the next step is to learn more about particular environmental job openings in each agency. There are a variety of ways this can be done. Visit a Federal Job Information Center, which has a complete listing of all government openings. Also, each agency produces a weekly listing of current job openings that are available at the agency's employment office, or by calling their job hotline. There are also private companies that compile job listing information and offer these lists in the form of periodicals. The subscription rate for these listings tends to be quite high, but most of these publications are available to you to browse through free of charge at many municipal and college libraries. Two publications that list a large number of federal job openings are:

Federal Career Opportunities. Published by Federal Research Service Inc., P.O. Box 1059, Vienna, VA 22183-1059. (703) 281-0200. This is a biweekly publication that lists thousands of federal job vacancies. It is organized by general service (GS) series within each agency. It also lists the name of the agency contact person as well as their telephone number. The subscription rate is $7.50 per issue or 6 issues for $38.00.

Federal Job Digest. Published biweekly by Breakthrough Publications Inc. It contains over 3,000 job vacancies in federal agencies throughout the country.

Another excellent way to find out about federal employment is to attend a Federal Job Fair. Job fairs are typically held in Washington, DC, and are advertised in the local newspapers. Call the employment office of any agency to find out when they are holding a job fair. These fairs are organized by either individual agencies or by a group of agencies with similar job openings. Job fairs are often conducted in large arenas or convention halls with scores of agencies and hundreds of job types to choose from.

There are information booths where you can speak directly with someone about job openings. In many cases, on-the-spot interviews are conducted, and if your credentials are impressive, agencies are often authorized to make an immediate job offer. You should therefore approach these fairs just as you would any job interview. Carry a stack of SF 171 forms with you and hand them out freely.

Additional Sources of Information

Almanac of Government Jobs and Careers (1991). By Ronald and Caryle
  Krannich and published by Impact Publications. Jobs and careers.
How to Get a Federal Job (1989). By David E. Waelde and published by
  Fedhelp Publications.

A Note on Government Bureaucracy

There are certain advantages and disadvantages to working in government. First, we are all familiar with its reputation as a sluggish and ineffective bureaucracy. One must beware that government agencies tend to foster large, impersonal work atmospheres where one can easily get lost in the bureaucratic shuffle. The career-minded environmentalist should be cautious of dead-end jobs and the "old boy" network that could block your career advancement. In recent years, government agencies have been working hard to dispel their bureaucratic reputation. Management techniques, which bring everyone into the decision-making process and encourage innovative thinking, have been borrowed from the private sector and are working.

One's work, especially anything having to do with environmental regulations and policy, is prone to the shifts in political tides. The Reagan administration, with its pro-business, pro-growth doctrine, was extremely hostile toward all federal environmental activities. In the early 1980s the administration attempted to gut the regulatory powers of the EPA. As a result, many EPA employees felt that their efforts were being ignored. This created a serious morale problem in the agency. With the present political atmosphere being less antagonistic toward environmental activities, there is much less of this type of pressure.

Environmental Jobs in Major Federal Departments and Agencies

The following is only a partial listing of jobs for career-minded environmentalists in the federal government. While there are many other departments, agencies, and offices that hire environmental professionals, these have the heaviest concentrations of environmental jobs and therefore, are described in detail.

The Environmental Protection Agency

The EPA was created in 1970 as an independent agency to consolidate the environmental activities of five executive departments and various other agencies. Its basic purpose is to carry out federal laws to protect the environment, especially in the areas of clean air and water. The EPA is responsible for the enforcement of most federal environment laws, which gives it a full agenda. The EPA has grown tremendously since its inception and is now the largest regulatory agency in the U.S. government in terms of budget and personnel. In 1970, the EPA employed 5,400 people with an operating budget of $900 million, while today it has 17,000 employees and a budget of more than $6 billion! There are 5,700 people working at the EPA's headquarters in Washington, DC, which serves as its administrative center, and there are an additional 11,000 environmental professionals employed at the EPA's ten regional offices. These offices are located in Boston, New York, Philadelphia, Atlanta,

Chicago, Dallas, Kansas City, Denver, San Francisco, and Seattle. The regional offices are responsible for carrying out and
enforcing all federal environmental laws and regulations. The
regional personnel work directly with state and local agencies,
industry, organizations, and individuals.

The EPA is organized into nine divisions that deal with specific environmental areas. These programs are: administration
and resource management; enforcement and compliance; policy
planning and evaluation; air and radiation; water; pesticides and
toxic substances; solid waste and emergency response; international activities; and research and development. In addition,
there are 25 EPA scientific research facilities located throughout
the nation.

The EPA hires about 1,000 new employees annually—most of
these openings are at entry level positions (GS-5, 7, and 9). A
little more than half of all new employees have some type of
scientific training. The most frequently advertised job opening
is for environmental engineers. The majority of the remaining
job openings are for environmental protection specialists. Career-minded environmentalists with backgrounds in political
science, sociology, geography, economics, community planning,
public policy, environmental studies, communications, and
journalism are all well represented in these positions. These
specialists do a great variety of work including developing regulations and policy, preparing and reviewing environmental impact statements, public relations, and consulting state and local
officials.

As head of the EPA under the Bush administration, William
Reilly, former president of the World Wildlife Fund and the
Conservation Fund, has gotten the EPA back on track after years
of politically induced stagnation. Its scope has been broadened
to include global environmental concerns like ozone depletion
and global warming, and the agency has developed innovative
programming such as market incentive programs and a greater
call for voluntary action from industry to reduce pollution. The
EPA does little of the actual research and field work; it primarily

administers programs to environmental consultants who carry out the actual work.

Working for the EPA can be a rewarding but at times frustrating experience. It seems that the EPA has far fewer human and financial resources than it does environmental responsibilities. This seems to be the most common complaint echoed by EPA employees. Also, the EPA is being constantly criticized by both industry and environmentalists. This type of "no win" pressure has earned EPA employees the reputation of being rather thick-skinned to criticism. Regardless of the challenges and drawbacks, working for the EPA places you squarely at the forefront of national and global work on environmental issues. A career in the EPA is also a good stepping stone to jobs in the private sector. Many environmental consulting and engineering firms hire former EPA employees. The transition to the private sector often brings with it an increase in salary and a bump up the career ladder into positions with more responsibility and prestige.

The United States Department of Agriculture (USDA)

This department houses a large number of services that focus on environmental activities. The USDA has over 122,000 employees, which makes it the largest single employer of environmental professionals in the federal government. Its major functions include agricultural research, support to farmers, and management of the national forest system. The USDA works closely with state agricultural agencies, as well as foreign nations. The USDA has traditionally supported environmental values and is highly respected for environmental conservation practices throughout the world.

Employment inquiries for all of the following services can be made to: Staffing and Personnel Information Systems Staff, Office of Personnel, Department of Agriculture, Washington, DC 20250, phone (202) 426-3964.

The Forest Service

The Forest Service manages 144 million acres of public land, mostly in 156 national forests in 44 states, the Virgin Islands, and Puerto Rico. A guiding principle of the service is its formula for mixed land use between industrial and recreational interests. The service has been criticized, especially in the West, for inadequately managing the clear cutting of timber by logging companies. Recently the service adopted an innovative program for tree harvesting that focuses on maintaining species biodiversity. The ongoing Spotted Owl controversy in the Northwest has ensured that the Forest Service will be at the center of the battle between environmentalists and lumber interests for quite some time.

The Forest Service employs a wide range of career-minded environmentalists, including biologists, foresters, and arborists. Any of these backgrounds combined with experience or a degree in business management is advantageous. Forest rangers, who are responsible for the management of public recreational areas, need both a solid understanding of the physical sciences as well as the ability to interact with the public. The service also employs a number of seasonal summer workers and has a limited number of paid intern programs in the Job Corps and the Senior Community Service Employment Program.

The Agricultural Research Service

This is the research branch of the USDA. Some of its recent activities include research on plant and animal production and protection, problems with the distribution of farm products, human nutritional studies, and air, water, and soil conservation. The Agricultural Research Service has over 8,000 employees, and a large number of these individuals hold advanced degrees in the agricultural and biological sciences. Most of the service's activities are conducted in cooperation with state agencies. A large share of their research is conducted at state and private universities and their field experimental stations.

The Extension Service

This service is the educational arm of the USDA, and it works with state and local partners to form the national Cooperative Extension System. The Extension Service maintains a small staff in Washington, DC, which provides leadership in developing educational programs for rural American farmers. In addition, state and local cooperative members hire a number of professionals to administer these programs. The focus is on providing scientific knowledge to improve the quality and yields on rural farms. Current extension concerns include sustainable agriculture, reviving small farms, food safety and quality, and waste management.

The Cooperative Extension System hires individuals with backgrounds in the agricultural sciences—agronomy, biology, and animal sciences; agricultural economics; the food sciences; and natural resources. Most extension professionals have a master's or bachelor's degree with some substantive training and are hired through the individual state land grant universities (each state has at least one).

The Extension Service publishes a booklet entitled *Commitment to Change*, which lists addresses for all state and U.S. territory land grant universities. To get a free copy of this brochure write: USDA Extension Service, Rm. 3337 South Building, 14th and Independence Ave., Washington, DC 20250-0900.

Department of the Interior

Unlike the USDA, this department has earned a reputation of hostility toward conservation values, especially during the Reagan years. This is unfortunate because the Department of the Interior has the most far-reaching jurisdiction over national environmental issues. Its responsibilities include: the protection and management of more than 549 million acres of public land (about 28 percent of the land area in the United States); the protection and preservation of wildlife; management and conservation of wetlands; and the enforcement of federal surface mining regulations. The department has always been at the center of the battle between conservation and development, and it has historically sided with the latter. Reagan's appointment of James Watt as secretary threw the department into such controversy that it has softened its political bias toward industry in recent years.

The National Park Service

This service runs the nation's 350 national parks, historic sites, monuments, and recreation areas. Environmental professionals from a wide variety of backgrounds work to formulate and administer policies, maintain park lands, and educate the public. Backgrounds in forestry, business administration, geography, parks and recreation, and the physical sciences are well represented in this service. Park rangers, who function as both environmental educators and law enforcement officials, play an important role in the service. There are also many openings for summer seasonal employment. Applications for seasonal employment must be received between September 1 and January 15.

For more employment information contact: Division of Personnel Management, Department of the Interior, P.O. Box 37127, Washington, DC 20013, phone (202) 208-5093.

The U.S. Fish and Wildlife Service

This service is responsible for the management of the 465 national wildlife refuges and 150 waterfowl production areas encompassing more than 90 million acres. Its jurisdiction also includes areas in the national park system that are reserved for hunting and fishing. Its mission is to "preserve and enhance fish and wildlife and their habitats for the continuing benefit of the American people." Other responsibilities include the enforcement of wildlife laws, surveillance of pesticides, and the listing of endangered species. The service employs professionals with specific training in fields like fish and wildlife biology, wildlife management, limnology, toxicology, and taxonomy, as well as engineers and chemists.

Employment inquiries should be made to: Office of Personnel, U.S. Fish and Wildlife Service, Department of the Interior, Washington, DC 20240.

The U.S. Geological Survey (USGS)

The USGS was established in 1879 to provide for "the classification of public lands and the examination of geological structure, mineral resources, and products of the national domain." Its original project was to map the entire nation and make this information available to the public. Its present concerns focus on investigating natural hazards like earthquakes, volcanoes, landslides, floods, and droughts. It also examines the nation's mineral and water resources. The USGS is one of the most productive services in terms of the number of scientific studies that it publishes. Employees include geographers, cartographers, geologists, engineers, and others trained in the physical sciences.

All employment inquiries can be made to: Geological Survey, Recruitment and Placement, 215 National Center, 12201 Sunrise Valley Dr., Reston, VA 22092, phone (703) 648-6131.

The Bureau of Land Management (BLM)

The BLM oversees 270 million acres, which makes it the single largest federal manager of public property. Most of this land is located in the West and Alaska. Its central responsibility is to manage energy and mineral exploration by private companies. The bureau maintains a policy of multiple use and sustainable yield practices for its three principle interests: forestry, mining, and recreation. Its "coziness" with industry has earned it the reputation of being pro-development and quite unsympathetic to environmental concerns. The BLM primarily recruits individuals with training in the hard sciences like civil engineers, mineral and petroleum engineers, cartographers, and biologists. There are also a limited number of openings, primarily in the middle and senior level positions, for social scientists and those with administrative training.

Questions about employment should be directed to: Personnel Officer, Bureau of Land Management, Department of the Interior, Washington, DC 20240. They also publish a booklet entitled *Career Opportunities in the BLM*.

The Bureau of Reclamation

These people are considered the "busy beavers" of the federal government, because they have overseen the construction of most major dam projects in the West. In addition, the bureau provides water to towns, farms, and industry, oversees the generation of hydroelectric power, and provides river regulation and water control measures. The administration of water rights is no small chore since the semiarid West has grown tremendously in population and in acres used for farming and livestock in the last 20 years. The bureau has had a stormy relationship with environmental groups and has fought some historic battles with the Sierra Club over the building of the Hetch Hetchy and Hoover dams and the proposed dam at the head of the Grand Canyon. Thus, the atmosphere at this bureau is often politically charged.

The bureau primarily hires individuals with degrees in civil and mechanical engineering, geology, hydrology, and the soil sciences. Employment inquiries should be directed to: Personnel Office, Engineering and Research Center, Building 67, Denver Federal Center, Denver, CO 80225, phone (303) 236-6914.

Department of Commerce

The National Oceanic and Atmospheric Administration (NOAA)

This administration's mission is to map, chart, and explore the ocean and to monitor, describe, and predict conditions in the atmosphere and space environment. Americans are very familiar with two of NOAA's activities: it provides the public with daily weather forecasts and advises of any potential destructive natural events through the National Weather Service; and it makes available to the public a complete mapping of all the nation's navigable waterways. Presently, NOAA is assessing the consequences of accidental environmental modification (pollution) and its effects on the earth and its human population over several scales of time. To their credit, scientists at NOAA (along with NASA) discovered the ozone hole over Antarctica. NOAA is also involved in research on alternatives to ocean dumping and in the development of national policies for ocean mining and energy exploration. NOAA is well respected, both nationally and internationally, for its technical knowledge of environmental problems.

NOAA employs a wide range of professionals with expertise in meteorology, oceanography, cartography, geography, geology, mathematics, and physics. For employment information contact the Department of Commerce, Washington, DC 20230, phone (202) 377-2985.

Department of Labor

The Occupational Safety and Health Administration (OSHA)

OSHA was created in 1970 to enforce all federal safety and health regulations. Its ongoing job is to improve safety standards in the work place and improve work place health, maintain records on job-related injuries and illnesses, and monitor federal agency safety programs. In fact, it was OSHA that pushed for public disclosure on the dangers of asbestos. OSHA has done a good job but, like the EPA, it has had its troubles with recent presidential administrations.

OSHA has a number of openings for safety and occupational health specialists, industrial hygienists, and policy analysts. Filling these positions are career-minded environmentalists with backgrounds in the health sciences, including occupational health nurses, engineers, and social scientists. Further employment information can be requested from: OSHA, Personnel Office, N3308, 200 Constitution Ave. NW, Washington, DC 20210.

Department of Defense

The U.S. Army Corps of Engineers

The corps has an unsavory reputation among environmentalists because it has, in the past, vigorously pursued engineering projects on rivers, harbors, and waterways without taking into account the environmental damage to aquatic flora and fauna. Twenty years ago it straightened out the meandering Kissimee River in Florida to control seasonal flooding. This extensive canal and dam system has wreaked damage on the fragile surrounding wetlands and wildlife, and the government now will have to spend millions of dollars over the next several years to restore the river to its original meandering state. The corps is, however, responsible for the enforcement of the regulations pertaining to the nation's wetlands, which is a large and very important task. Over the past few years a battle has been brewing over the definition and designation of wetlands and the corps is squarely in the middle of this debate.

The corps is a branch of the U.S. Army but hires civilians for all types of engineering positions. Employment inquiries can be made to: Personnel and Employment Service–Washington, Room 3 D727, The Pentagon, Washington, DC 20310-6800.

The National Science Foundation (NSF)

The NSF awards grants and scholarships for research and education in all areas of science and engineering. In recent years, an increasing number of awards have been granted to programs and projects geared toward environmental problems. While no actual research is conducted by the NSF, its staff reviews and analyzes research proposals in such areas as engineering, agriculture, science, medicine, education, and public affairs. Thus, the NSF sets the national agenda for exploring scientific knowledge. Staff members work primarily in their areas of interest and gain a solid understanding of the latest research issues.

Since the NSF is involved in such a variety of scientific issues, it hires individuals with training in both the physical and social sciences. At the purely administrative level, aspiring professionals with degrees in management, accounting, and contract administration are solid candidates. For more employment information contact: Division of Personnel and Management, National Science Foundation, Room 208, Washington, DC 20550, phone (202) 357-9859.

The Office of Technology Assessment (OTA)

The OTA helps Congress explore complex issues involving science and technology and offers suggestions on national policy alternatives. It conducts in-depth studies for congressional committees and individual members, as well as more immediate congressional needs like testimonies, special reports, and briefings. The OTA operates the Science, Information, and Natural Resources Division, which explores a wide range of environmental issues. Staff members form teams that work closely with congressional staffs on specific projects. The OTA is well respected by congress, and its members are often exposed to the dynamic political atmosphere on Capitol Hill.

This office hires a small number of research assistants with a wide variety of college degrees. A job with the OTA is often a stepping stone between undergraduate and graduate studies. For further job information contact: Office of Technology Assessment, Personnel Department, United States Congress, Washington, DC 20510–8025.

The Congressional Research Service (CRS)

The CRS is the independent and nonpartisan research and reference arm of Congress. Its staff works directly and exclusively with individual members of Congress. Staff members conduct research, analyze policy issues, and answer questions for congressional members. Thus, the CRS is the heart and soul of the information flow through Congress. There are two divisions that deal with environmental issues: Science Policy Research and Environment and Natural Resources.

The service looks for individuals who excel in their academic field. Many staff positions are for policy analysts with backgrounds in sociology, economics, foreign affairs, biological sciences, engineering, and physical sciences. Reference librarians and paralegal assistants are also in demand. The CRS offers a limited number of unpaid volunteer positions for college students. For more employment information contact: Staffing Team, CRS Administration Office, James Madison Memorial Building, LM 208, The Library of Congress, Washington, DC 20540, phone (202) 707-8803.

The Council on Environmental Quality (CEQ)

This council was created to formulate and promote national policies to improve the quality of the environment and report their findings directly to the president. In recent years, the council has been called upon infrequently, but past administrations with more sound environmental priorities have used many of its policy recommendations. The council prepares the *Annual Environmental Quality Report* for the president, appraises federal programs to determine if they promote sound environmental policy, and oversees the implementation of the National Environmental Policy Act.

This office also hires a small number of researchers with varied college training. For more employment information contact:

Executive Office of the President, Personnel, 725 17th St. NW, Office of Administration, Room 4013, New Executive Office Building, Washington, DC 20503.

The Peace Corps

The Peace Corps was created to promote world peace and friendship through the helping activities of its volunteers. These volunteers are sent to over 70 countries throughout the world to help solve development problems. This work is conducted through the following six program areas: agriculture; education; health; urban development; small business development; and the environment. This is a wonderful way to learn new customs, languages, and technical skills, as well as traditional practices and life-styles that have a low impact on the environment. Volunteers go abroad for two years and live in the communities in which they work. There is a small stipend paid to volunteers at the end of their service.

In addition to volunteer positions, there are jobs for those interested in recruitment and managerial careers. For information on becoming a volunteer contact: Peace Corps, Office of Volunteer Services, Washington, DC 20526, phone (202) 606-3336. For those interested in working in an administrative position contact: Peace Corps, Office of Personnel Management, Washington, DC 20526, phone (202) 775-2214.

Environmental Opportunities in Other Federal Departments

There are numerous other places in the federal government for the career-minded environmentalist. In the Department of Transportation try the Federal Aviation Administration, Federal Highway Administration, Materials Transportation Bureau, and the National Transportation Safety Board. At the independent commissions there are environmental jobs at the Consumer Product Safety Board, Federal Energy Regulatory Commission, Federal Maritime Commission, and the Federal Trade Commission.

Jobs in Congress

There are numerous job opportunities for the career-minded environmentalist on Capitol Hill. The primary reason for this is that there is a 40 percent, or 4 in 10, staff turnover rate per year. The work atmosphere on the Hill is much different than that in most other parts of the government. The work pace is on the boom or bust cycle, where staffers work feverishly for days on end, followed by a few days of calm before the next crisis. Congressional staffers are the true movers and shakers in the halls of Congress for they do a lion's share of the work and receive little if any of the credit. Career-seeking environmentalists who have a flair for the fast-paced political game and the fortitude to work long hours and receive little pay, will find the Hill a rewarding experience. While few staffers stay on the Hill for more than a year or two, the experience and connections made there more than compensate for the drawbacks.

A Brief Tour Through Congress

Congress is made up of the House of Representatives and the Senate. Each state has two senators while the number of representatives varies depending on state population. Vermont, for example, has one congressperson while California as over 40. Staff workers in Congress are *not* federal employees, meaning that you don't have to fill out the dreaded SF 171 to get a job. In fact, Congress, the law-making branch of the federal government, has made itself exempt from most federal hiring and salary regulations. Thus, the job search process in Congress is more like that of the private sector, because there are few if any bureaucratic procedures for congressional offices to follow.

Each congressperson maintains two sets of staff: one in their home district and one in Washington, DC. Their Washington offices have the larger and more politically active staff. You should concentrate your search on congressional members (most of these are Democrats) who have a reputation as environmentalists. Other members of Congress are deeply involved in environmental issues through their work on the 25 committees in both the House and Senate that have an environmental agenda. Figure 2 is a list of the committees and subcommittees that focus on these environmental issues.

Committee staff members do not work for specific congressional members, but instead work for a committee that deals with specific pieces of environmental legislation. Contact the congressional committees directly for job openings.

Sources of Information

The Almanac of American Politics (annual). By Michael Barone and Grant Ujifusa and published by the National Journal. This book gives a full profile on individual members of Congress, their voting records, and how they are rated on such issues as the environment, defense, and economic policy. This book is touted as being one of the six books that Jimmy Carter displayed prominently on his White House desk.

The Congressional Staff Directory (annual). Available from Charles Brownson (a former member of Congress) P.O. Box 62, Mt. Vernon, VA 22121. This book lists the names of every congressional staffer, as well as specific biographical information like their academic and employment history and membership in organizations.

Capitol Jobs: An Insiders Guide to Finding a Job in Congress (1986). By Kerry Dumbaugh and Gary Serots and published by Tilden Press. An excellent general guide to understanding the ins and outs of getting a job in Congress.

Congressional Careers: Contours of Life in the U.S. House of Representatives (1991). By John R. Hibbling and published by the University of North Carolina Press.

Environmental Careers in State Government

In the last ten years, state governments have begun to play a major role in the development of environmental policies and

Figure 2   Congressional Committees and Their Jurisdictions
The following are the congressional committees with jurisdiction over environmentally related programs and the programs for which each committee is responsible.

SENATE

Agriculture
Soil conservation, ground water
Forestry, private forest reserves
Pesticides, food safety
Global change

Appropriations
International monetary and financial funds
Forest Service
Army Corps of Engineers
Nuclear Regulatory Commission
Tennessee Valley Authority
Occupational Safety and Health Administration (OSHA)
Mine Safety and Health Administration
Soil conservation programs
Food and Drug Administration
Environmental Protection Agency
Council on Environmental Policy
National Oceanic and Atmospheric Administration

Armed Services
Military weapons plants
Nuclear energy
Naval petroleum, oil shale reserves
Air Force jet emissions

Commerce, Science, Transportation
Coastal zone management
Inland waterways
Marine fisheries
Oceans, weather, science research
Outer continental shelf
Global change

Energy and Natural Resources
Energy policy, conservation
National parks, wilderness
Nuclear energy, public utilities
Public lands, forests
Global change

Environment and Public Works
Environmental policy, oversight
Air and water pollution
Outer continental shelf
Toxic substances
Fisheries and wildlife
Flood control, deep-water ports
Ocean dumping
Nuclear energy
Bridges, dams, inland waterways
Solid-waste disposal
Superfund, hazardous waste
Global change

Finance
Revenue measures, user fees

(Figure 2 continued)

Foreign Relations
Nuclear energy, international
International Monetary Fund
International environmental
   affairs

Government Affairs
Nuclear export policy
Nuclear weapons plant cleanup

Judiciary
Environmental law, penalties

Labor and Human Resources
Occupational health and safety
Pesticides, food safety

HOUSE

Agriculture
Agriculture and industrial
   chemistry
Soil conservation, ground water
Forestry and private forest reserves
Pesticides, food safety
Global change

Appropriations
International monetary funds
Forest Service
Army Corps of Engineers
Nuclear Regulatory Commission
Tennessee Valley Authority
Occupational Safety and Health
   Administration (OSHA)
Mine Safety and Health
   Administration
Soil conservation programs
Food and Drug Administration
Environmental Protection Agency
Council on Environmental Policy
National Oceanic and
   Atmospheric Administration

Armed Services
Military weapons plants
Naval petroleum, oil shale reserves
Nuclear energy

**Banking, Finance and Urban
Affairs**
International financial and
   monetary organizations

Education and Labor
Occupational health and safety

Energy and Commerce
Energy policy, oversight
Energy conservation
Health and the environment
Interstate energy compacts
Public health and quarantine
Nuclear facilities
Transportation of hazardous
   materials
Solid-, hazardous-waste
   disposal

Foreign Affairs
Foreign loans, IMF
International environmental
   affairs
Global change

Government Operations
Environment, energy, natural
   resources oversight

(Figure 2 continued)

Interior and Insular Affairs
Forest reserves (public domain)
Public lands
Irrigation and reclamation
Petroleum conservation (public lands)
Conservation of radium supply
Nuclear energy industry

Judiciary
Environmental law, penalties

Merchant Marine and Fisheries
Coastal zone management
Fisheries and wildlife
Oil-spill liability
Wetlands

Public Works and Transportation
Flood control, rivers and harbors
Pollution of navigable waters
Bridges and dams
Superfund, hazardous waste

Science, Space and Technology
Research and development
National Weather Service
Global change
Nuclear energy, facilities
Agriculture research
National Oceanic and Atmospheric Administration

Ways and Means
Revenue measures, user fees
Superfund

Source: Phillip Marwill, *Congressional Quarterly Weekly Report*, January 20, 1990, 151.

regulations. While up to the early 1980s the federal government took the lead environmental role, the past decade has seen a dip in federal agenda setting and a rise in state activities. California is frequently cited as the state leader in pollution abatement measures, and it appears that other states are also beginning to realize that they need to aggressively address their own environmental problems. Serious issues like air pollution, urban sprawl, hazardous waste disposal, and drinking water contamination have forced the states to take actions that in many cases go beyond regulations set by the federal government.

State environmental agencies expend most of their energy carrying out specific programs and distributing information, funding, and resources to county and local governments. While each state does not have the number or breadth of environmental departments and agencies as the federal government, they do offer a wide variety of job opportunities for the career-minded environmentalist. Also, state environmental specialists are more apt to work on projects from "cradle to grave" and have more input than their federal counterparts. While resources may be more limited at the state level (some states are much more limited than others), these agencies are smaller and tend to be less bureaucratic.

While the number and types of jobs vary widely, each state has its own environmental protection department. This is a good place to start when researching for career opportunities. Appendix B is a listing of state environmental agencies, their addresses, and phone numbers. Appendix C is a listing of offices dealing with air pollution, land use, solid waste, and radiation. Also, Figure 3 shows the number of environmental department employees and the type of jobs by state.

Sources of Information

Opportunities in State Government. Published by ACCESS, 50 Beacon St., Boston, MA 02108, (617) 720-5627.

Directory of State Environmental Agencies (annual). Published by the Environmental Law Institute, 1616 P St. N.W., No. 200, Washington, DC 20036.

Careers in State and Local Government (1980). By John William Zehring and published by Garrett Park Press.

Environmental Careers in Local Government

In local government, the career-minded environmentalist will find a potpourri of jobs. The most common denominator for these jobs is the emphasis on hands-on work. A good chunk of the money that federal and state governments allocate for environmental projects is carried out by city, county, and municipal authorities. Water and waste treatment plants, landfills, garbage collection, and recycling programs are some of the projects that local governments operate and/or regulate. These are essential

Figure 3 Environmental Department Personnel by State

State	Authorized positions	Actual employees	Permit reviewers	Inspectors	Laboratory personnel	Enforcement personnel	Technicians/ field data	Policy analysts	Administrators
Alabama	260	245							
Alaska	250	215							
Arizona					22			32	
Arkansas	217	170							
California	583	586	14	23	88	56	185	110	110
Colorado[a]	1,247								
Connecticut	1,947	1,618	127	50	1	213	180	38	202
Delaware									
Florida	1,222	1,106							
Georgia									
Hawaii	302	278	28	32	0	27	134	45	36
Idaho									
Illinois	834	770							
Indiana									
Iowa									
Kansas	255	230	58	50	20	5	37	c	45
Kentucky	1,322	1,239	147	295	22	294	51	45	385
Louisiana	396	356	36	175	17	5		26	91
Maine									
Maryland	682	639	92	214	0	28	126	38	184
Massachusetts									
Michigan	671	619	84	170	42	23	80	110	110
Minnesota									
Mississippi	749	621							
Missouri	446	413	31	92	13	33	63	27	132
Montana	112	103	14	24	19	4	15	11	18
Nebraska	114	102	12	16	6	3	7	25	32

(Continued)

(Figure 3 continued)

State	Authorized positions	Actual employees	Permit reviewers	Inspectors	Laboratory personnel	Enforcement personnel	Technicians/ field data	Policy analysts	Administrators
Nevada	46	44	14	10	0	4	4	5	9
New Hampshire									
New Jersey	3,294	3,184							
New Mexico	301	264	44	65	0	37	49	16	90
New York	3,647								
North Carolina	1,931	1,931	25	37	0	12	20	18	24
North Dakota	84	80							
Ohio	700	660							
Oklahoma	1,063		301	558	43	29		26	164
Oregon	387	310	50	92	63	5	37	10	140
Pennsylvania	3,756	3,518							
Rhode Island		489							
South Carolina	477		55	60	50	25	120	58	109
South Dakota									
Tennessee	815	790	56	416	0	28	81	59	150
Texas	351	332	28	104	45	10	38	23	103
Utah[b]	1,014	1,014							
Vermont	517	552	43	67	22	7	153	73	131
Virginia									
Washington	796	750	73	64	30	36	123	94	255
West Virginia	60	60	4	22	0	0	0	9	19
Wisconsin	2,595	2,409							
Wyoming	151	139		83	5		6	6	49
TOTALS	32,531	26,899	1,336	2,719	508	884	1,509	904	2,588

Source: R. Steven Brown and E. Garner, Resource Guide to State Environmental Management (Lexington, Ky.: The Council of State Governments, 1988).
Notes: Departmental duties vary considerably from state to state. The data should not be interpreted as an indicator of a state's commitment to environmental programs. Blanks indicate not applicable or data not available. In some cases the actual employees column does not equal the breakdowns in the other seven columns. Totals may vary for a variety of reasons: one position may be listed in more than one category because of job responsibilities or some positions do not fit in one of the seven categories listed.
[a]Department of Natural Resources. [b]Department of Natural Resources and Energy. [c]Included with administrators.

services and they must be continuously maintained and up-graded.

While federal and state environmental agencies have passed a larger share of responsibility on to local governments in recent years, an increasing number of citizen groups have also been pressuring for better monitoring of community problems. The combination of these demands have made local authorities more responsive to community needs.

Working at the local level is a great way to begin an environmental career. There is a high turnover rate that translates into a steady flow of job openings. Jobs tend to be oriented toward field operations like program management, inspections, and enforcement, which are the building block skills that every career-minded environmentalist should learn. Also, the work tends to be decentralized, and new employees are quickly given responsibility that teaches valuable leadership skills. Since most work is carried out in small communities or urban neighborhoods, professionals constantly interact with citizens and feel much less a part of the government bureaucracy. There are some drawbacks, like relatively low pay compared to other government workers and a much greater degree of petty politics, but these jobs are a conduit to environmental careers in other parts of government and the private sector.

To find out about these jobs, contact local agencies directly. Look in the local phone book to find the department headings. Many public works and parks and recreation departments offer summer jobs to high school and college students. Also, ask neighbors or friends who may know someone employed in one of these departments, or a local politician who may be able to pull a few strings.

F I V E

Environmental Entrepreneurs

New Opportunities
for the Eco-entrepreneur

Many new and wonderful opportunities for the individual with an entrepreneurial spirit have developed as a result of the consumer's increased concern and awareness about the environment. A large and growing market has emerged for products that have been developed, packaged, and marketed with the least possible amount of harm being done to our natural resources, our health, the health of the workers, and the ozone layer. In addition, a new business ethic that has a respect for the environment at its foundation has emerged and is continuing to develop. In Chapter 3, we discussed how traditional jobs are now evolving to include an environmental component. In this chapter, we are discussing new opportunities for businesses that have been conceptualized and developed with environmental concerns in mind.

The desire to own a business, to make money, and to provide useful goods and services no longer is in conflict with good environmental policies. Past realities and misconceptions that green products and services were too expensive for widespread use are diminishing as more and more research and development is done. New, less expensive technologies and large-scale production capabilities are helping to make environmentally sensitive products less expensive and more accessible to the masses.

Another important result of the increased demand for green products is that the gap has begun to close between the grassroots environmentalist (and all the stereotypes that label carries with it) and the business world (and all the stereotypes it carries). Environmentalists have become more business minded and businesses have become more environmentally minded. This trend will continue as our concern for the environment grows, and it eventually will become the norm rather than the exception to the rule.

The opportunities are endless for the entrepreneur. A commitment to the environment coupled with a good idea are two of the essential components for beginning your career as an eco-entrepreneur. In the following pages we will provide an overview of some of the necessary qualifications for starting your own business and then outline the areas where the opportunities are presently the greatest for the entrepreneur. In addition we have provided names and addresses of organizations and resources that will be useful as you investigate further the possibility of starting you own business.

What Does It Take to Start Your Own Environmental Business?

Starting your own environmental business requires a lot of hard work and a commitment to blend the desire to make a profit with a strong sense of social responsibility. Consumers today are choosing products that make claims of being environmentally safe over those that do not. Very often these products cost more and sometimes are less convenient (bottles need to be saved, returned, recycled, and refilled) than the modern day consumer is used to. In order to successfully launch an environmental business or product, a mission statement or overall philosophy of the company needs to be articulated and presented to the buying public.

As you begin formulating the plan for your business you need to think about the company's philosophy, write it down, and develop it so that it can be expanded and adapted for every area of your business—from product development to in-house environmental policies. As more and more companies make claims about the "greenness" of their products and services, consumers are beginning to demand real information on the quality and commitment of the products, and the companies that produce them.

Qualifications

Three areas in which you will need to have well articulated plans and policies are environmental ethics, recycling programs, and research and development. Clearly defining your goals and commitment in these areas will give you the foundation you need to develop a successful plan for your business. The type and size of your business will help you formulate the philosophy of the organization. However, all of your environmental policies and plans should be flexible enough to grow and adapt with the growth of your business.

Environmental Ethics

A clearly defined commitment to environmental issues and to the protection of our natural resources is one of the most important criteria for consumers in choosing an environmentally safe product over one that is not. An international reliance on rural cultures for raw materials requires that the ethic of the company be one of mutual respect for and investment in social programs in the countries and communities that are providing you with the raw materials to produce your product.

This commitment begins early and could be as simple as donating a certain amount of your profits to social programs or environmental organizations. As the company grows it should follow that the company's involvement and commitment to social programs also increases.

In addition providing your employees with an environmental work ethic will guarantee that the entire organization is dedicated to providing useful goods and services that have the good of the earth in mind.

Recycling Programs

An important measure for monitoring a company's commitment to operating its businesses in an environmental manner is how trash is dealt with. This includes in-house policies for recycling and packaging the products you produce. An important question to address is what happens to product containers and packaging when consumers are finished with them?

Plans need to be made from the start to use minimal and recyclable packaging, and systems need to be set up to make it as easy as possible for the customer to return, refill, and recycle packaging.

Research and Development Contributions

Consumers are also interested in knowing about how much effort companies are investing in research and development of new products that further promote the protection of our natural resources. Many companies have their own labs that work continually to develop and test new products that don't exploit our natural resources and that are renewable. Consumers are interested in knowing just how committed you are to continued investment in products designed to save the earth.

Market Environment

In order to start a successful business, it is important to have more than just a good idea or a nice product. Before launching and developing a new product, you need to identify your market and do research to find out if consumers share your desire to invest in the future of the planet.

The amount and type of research you do will depend on the size and scope of your business. The simplest way to investigate how receptive the market will be to your product is to examine your own consumer behavior and that of your friends. Spend time in stores that are promoting environmental products and read magazines, books, and anything else that focuses on environmental concerns. If the product or service you are planning to start is larger than a basement operation and if you have the resources to hire market researchers to help you define and understand your market, you should do so. Talking and listening to your potential consumers are also an excellent and essential way to get a feel for the issues you will face and the concerns you will encounter.

Financing

Financing a new business—environmental or not—is a key element to the success of any new business venture. Before launching a new business, it will be necessary to investigate the ways to finance your efforts. Traditional banks are now seriously considering the long-term benefits of investing in environmentally sound companies, and as a result, are much more receptive to investing in companies that have the preservation of our natural resources as a fundamental tenet of their corporate philosophy. When applying for a loan, investigate which banks have already invested in environmental companies, and talk to other eco-entrepreneurs to get information on where they turned to get financial support.

In addition, the Global Environment Fund (GEF), founded in 1989, invests in companies whose products and services contribute directly to environmental improvement. As of mid-1991, the fund had over $20 million dollars in assets, with over 60 growth-oriented companies in its portfolio. The GEF looks for innovative companies that share its determination to take on the challenge of working for the good of the environment by creating and supporting new businesses that provide services, better techniques and technologies, more efficient production processes, and new products to replace environmentally harmful ones. Environmental correctness is not the only criteria used to allocate funds, however. The GEF also looks for companies that are financially savvy, demonstrating an ability to achieve a superior rate of return and an understanding of the importance of enhancing shareholder value. The qualities that the GEF looks for once a company has been identified as well positioned to take advantage of a specific environmentally oriented business opportunity include the following.

1. A well-articulated business plan. This means that the entrepreneur has an understanding of the current market position and a promise for substantial long-term growth.

2. Advanced product development and commercialization. Because new environmentally sensitive technologies are being developed, the GEF looks for people who are using and adapting existing technologies to deliver better services.

3. Unencumbered by previous financial arrangements. The GEF especially looks for companies that are not constrained by previous rounds of debt or venture capital financing.

4. Strong and proven management teams. The structure of the company needs to be clearly defined with a well-qualified and experienced management team in place.

5. Products or technologies with multiple applications. Companies that can do a variety of things or provide a variety of services are considered to be the best investments.

6. Strong profit margins. New businesses need to consider how their environmental products, which often cost more than traditional nonenvironmental products, can compete and still be profitable.

For more information on the Global Environment Fund you may wish to contact them directly.

Global Environment Fund, L.P. GEF Management
   Corporation
1250 24th Street, NW
Suite 300
Washington, DC, 20037
(202) 466-6454

Overview of Entrepreneurial Opportunities

The following is a list of areas where opportunities are plentiful for the ecoentrepreneur. The categories outlined are broad and many different types of businesses could fall into each category. This list should give you an idea of where the opportunities lie as you begin thinking about your own business plans.

Renewable Energy

Increased public concern about diminishing and expensive resources that provide us with heat, fuel, and the comforts of our day-to-day lives has resulted in a new interest in renewable energies. People with ideas and inventions that promote the use of solar, wind, and other renewable energy sources are becoming more and more in demand. From electric cars to rechargeable batteries to solar homes, the opportunities here are almost limitless.

Recycling

As recycling becomes more and more a part of our everyday lives, new opportunities are emerging. The development of environmentally safe packaging alternatives and the collection and redevelopment of existing materials are two areas where ambitious entrepreneurs can make a difference and a living.

Retail

In response to increased concern about the environment and personal health and well-being, traditional markets that have not taken the environment into consideration are now being forced to do so. From cosmetics to clothing to cleaning products, new markets are emerging for environmentally sensitive products. The naturalness and simplicity of these products holds a great deal of appeal for the consumer who has a heightened awareness of the precarious state of our earth. Many mail order businesses have emerged, offering products that have not yet reached the shelves of our department stores. Canvas bags, recycling bins, and 100 percent cotton clothing are just a few of the items now being promoted and sold by eco-entrepreneurs.

Agriculture

A natural outgrowth of the environmental movement is an increased demand by consumers for foods that are preservative and pesticide free. New opportunities have emerged for producers of organic produce and healthy foods. In addition, products developed from renewable natural resources are being well received in the marketplace.

Construction

An integral part of the recycling movement is in the areas of construction and architecture. New buildings and furniture are more commonly being built from existing, used raw materials.

The opportunities here are endless and the financial rewards substantial.

Inventions

Working with existing, ineffective technologies to produce new methods and tools for producing and recycling products is a fast growing area for the environmental entrepreneur. In addition, new products are being developed with the ultimate goal of replacing existing products with new environmentally sensitive products. A good example of this is in the auto industry where an increased amount of time and money is being put into the development of electric cars.

Creative Careers

Many other opportunities exist for creative types who carry out their work with the environment in mind. Artists are using recycled materials in their artwork, and writers are writing books on the environment. People are selling goods that they have developed to meet their own needs, and good environmental products are selling.

As you can see, opportunities for the entrepreneur are plentiful. Each of these categories has within it profiles of a diverse group of people with varying skills and educational backgrounds. Some of these entrepreneurial ventures require a great deal of technical expertise and some just require a commitment to the environment and a good idea. If you choose to pursue a career as a entrepreneur, it will be important for you to think carefully about your own skills and interests and to combine them to develop a business that will be uniquely your own.

As with many other careers outlined in this book, this type of career requires that you possess many skills. Communication skills, an in-depth knowledge of environmental issues, and business training will help make your business a success.

The Eco-entrepreneurial Network

Integral to the success of your business will be how well you tap into the emerging network and resources available to people like yourself. Magazines, conventions, and organizations exist specifically for environmental entrepreneurs and will be extremely helpful to you as you develop, plan, and launch your own business.

Publications

The following is a brief list of some of the resources available to the ecoentrepreneur. This list is in no way complete but these publications, in addition to providing you with helpful information, will lead you to additional resources.

In Business
The Magazine for Environmental Entrepreneuring
JG Press, Inc.
419 State Avenue
Emmaus, PA 18049

Business Ethics Magazine
1107 Hazeltine Blvd.
Suite 530
Chaska, MN 55318

The Consumer Guide to Planet Earth
Schultz Communications
9412 Admiral Nimitz, NE
Albuquerque, NM 87111
(505) 822-8222

The 1992 Natural Connection Newsletter
(a directory of manufacturers and suppliers of environmentally
friendly and recycled products)
P.O. Box 8233 N.
Brattleboro, VT 05304

Conventions

In response to the increased number of businesses that focus on
environmentally safe products, the number of conferences,
expos, and meetings have increased. Your best source for finding
out about these events is by reading and subscribing to environmental publications. The National Marketplace for the Environment sponsors expos in Los Angeles, San Francisco, and Boston.
The 1992 ECO EXPO in Los Angeles had more than 375 exhib-
itors and 33,000 attendees. For further information about dates
and locations call (818) 906-2700.

Networking

Your most valuable resource, as with any career path you choose
to pursue, is yourself and others who are doing similar work.
Networking will be an integral part of your success in any career
you choose to pursue but it will be especially helpful if you are
choosing to start a business. Eco-entrepreneurs are interested in
the successful development of companies like their own and are
usually very willing to give advice and information to help you
get off to the right start and avoid making the same mistakes they
may have made.

SIX

Career Opportunities
in the Nonprofit Sector

The great American spirit of helping is once again glimmering as we move through the 1990s into an era of sustained environmental awareness. Of the many challenges that confront our nation and the world, none is more pressing than the preservation of our fragile environment, and no people have risen to the challenge better than the concerned Americans. Since the early 1980s, participation in the environmental movement has soared. Membership in most national environmental groups like the Sierra Club and the National Audubon Society has doubled or tripled, and thousands of new grassroots (local) environmental organizations have sprung up. The maxim "think globally, act locally" is flourishing as communities around the United States are fighting (and winning!) battles against toxic dumping and pollution, and are working together for the health of their families and fellow citizens. Americans are feeling a renewed sense of political and social empowerment and they are channeling this into environmental activities. The goal is to do something positive for our future, the future of our children, and the earth upon which we all vitally depend. We may be on the verge of a whole new political era since the election of this country's first potentially "green" presidential team, led by Vice-President Albert Gore, an unbashful and proven environmentalist.

In this chapter we will describe opportunities in national, state, and local nonprofit environmental organizations. Of all the career areas discussed in this book, there are actually the fewest job opportunities with environmental organizations. Although the number of new organizations and the level of membership and donations for existing organizations has grown tremendously, only the very largest have professional staffs of more than 50 people, while the majority of organizations have perhaps a handful of paid staff members or all volunteers with no paid staff at all. These organizations are perpetually short of money, and are often locked in legal and policy battles with large corporations that have far superior financial and human resources. This is compounded by the fact that the nonprofit sector still lags behind both government and the private sector in wages, benefits, and job security.

So, given these facts, why would anyone be interested in pursuing a career with a nonprofit environmental organization? The answer, oddly enough, is quite simple: these people know that they are making a measurable difference in our world through their environmental work. They are working in an exciting and challenging atmosphere that gives them the opportunity to meet hundreds or even thousands of like-minded and highly motivated individuals. In no other job sector are workers more dedicated or determined; many of these individuals view themselves as being like David battling the powerful Goliath, and they relish the opportunity to lay the big bully down to defeat!

The Nonprofit World

In these pages we will describe to you the rich opportunities and the ways in which you can do something positive for the earth by working with nonprofit environmental organizations. We will describe, in detail, the nine largest organizations to give you a feel for what they do and how they operate. We will also describe volunteer and intern positions in these groups. Keep in mind that state and local organizations work with or receive information and funding from these national groups, therefore, these organizations are good contacts to get information on career opportunities in and around your community. There are regional environmental centers that have listings of volunteer, internships, and job openings that will be explored. In addition, we will describe two environmental job placement organizations that specialize in finding jobs for career-minded environmentalists. Also, magazines, computer networks, and databases, which are all good resources to use in your job search, will be listed.

Sources of Information

Regional, State, and Local Organizations (1989). Published by Gale Research. This is an exhaustive listing of nonprofit organizations in the United States.

Your Resource Guide to Environmental Organizations (1991). By John Seredich (ed.) and published by Smiling Dolphin. Lists federal and state organizations and nongovernment organizations.

World Directory of Environmental Organizations (1991). By the California Institute of Public Affairs, Sacramento, CA: Sierra Club.

World Environmental Directory (annual). By Business Publishers, Inc. Including 13 entries with 30,000 listings of conservation and environmental organizations, pollution control product manufacturers and consultants, federal and state governments, attorneys with environmental interests, environmental education programs, databases, and funding sources.

National Environmental Organizations

Among these are some well-known organizations like the Sierra Club and the National Wildlife Federation. Most of these organizations make their headquarters in Washington, DC, and have regional offices and/or local chapters throughout the country.

These nonprofits serve a dual purpose: they encourage citizen participation at the local level, and they work to influence politicians and create environmental laws at the national level. Since the early 1980s, when membership and funding for many of the established environmental nonprofits soared, they have revealed themselves as a strong political voice in Washington.

There are far more intern and volunteer opportunities than staff positions at most of these organizations, and the best way to land a job is to first do volunteer or intern work. In fact, most job openings are not even advertised because they are given directly to interns and volunteers. By donating your time, you will get a first-hand opportunity to see how these organizations operate, and you can decide if a career in the nonprofit sector is for you. In many cases, intern and volunteer staff will try to match your interests and skills with projects, but be prepared for a good amount of busy work like envelope stuffing and bookkeeping.

The type of personnel hired depends largely on the organization and the issues that they pursue, although some generalizations are possible. Good communication and writing skills are essential. Most of these organizations are advocacy groups and, therefore, depend upon their employees to maintain open channels of communication with the public. Among the types of professionals hired are conservation specialists, journalists, policy researchers, economists, lawyers, and health specialists. Also, a good number of the national headquarter staff are administrators and fundraisers. Individuals with backgrounds in management, finance, and accounting, as well as individuals with fundraising experience and secretarial skills, are always in demand. The following is a list of some of the largest nonprofit environmental organizations in the country. Listed also are the number of supporters and staff members for each organization.

The Nature Conservancy is an organization whose aim is to "find, protect, and maintain the earth's rare species and natural communities by preserving the lands they need to survive." To achieve this goal, the Nature Conservancy locates ecologically

Leading Nonprofit Environmental Organizations in the United States

Organization	Staff	Members/Supporters
Environmental Defense Fund 257 Park Avenue South New York, NY 10010 (212) 505-2100	145	200,000
Greenpeace USA 1436 U Street NW Washington, DC 20009 (202) 462-1177	200	1,800,000
National Audubon Society 950 Third Avenue New York, NY 10022 (212) 832-3200	320	600,000
National Wildlife Federation 1400 16th Street NW Washington, DC 20036 (202) 797-6800	650	5,300,000
Natural Resource Defense Council 40 West 20th Street New York, NY 10011 (212) 727-2700	160	170,000
Nature Conservancy 1815 North Lynn Street Arlington, VA 22209 (703) 841-5300	1,400	672,000
Sierra Club 730 Polk Street San Francisco, CA 94109 (415) 776-2211	328	570,000
Wilderness Society 900 Seventeenth Street NW Washington, DC 20006-2596 (202) 833-2300	135	317,000
World Wildlife Fund—U.S. 1250 24th Street NW Washington, DC 20037 (202) 293-4800	300	1,000,000

sensitive or threatened lands, acquires these lands, and then maintains them as nature preserves. Presently, the Conservancy manages nearly 1,440 preserves on over 6 million acres of land throughout the United States, Canada, and Latin America. The Conservancy has entered agreements with other nonprofit organizations and the U.S. government to greatly expand these holdings over the next decade. The Latin American debt-for-nature swap, where governments are forgiven a part of their foreign debt in exchange for creating national nature preserves is among the Conservancy's greatest achievements.

The Nature Conservancy's headquarters are in Arlington, Virginia, which is located just outside of Washington, DC. There are also four regional offices in Boston, Massachusetts; Minneapolis, Minnesota; Chapel Hill, North Carolina; and Boulder, Colorado, as well as local offices in every state. There are presently about 1,400 professionals working for the Conservancy, which makes it the largest nonprofit environmental employer. Most of these individuals work on the nature preserves and have degrees or experience in resource and wildlife management. Volunteers are an integral part of the Conservancy's success and are encouraged to take an active role in managing and maintaining the preserves and helping out at the local offices. Call the main or regional offices for the location of the local office or nature preserve nearest to you.

The National Wildlife Federation (NWF), which was created with the help of Franklin D. Roosevelt in 1936, is involved with conservation efforts in the United States and throughout the world. In terms of membership, the NWF is the largest nonprofit environmental organization in the United States, with 5,300,000 members. The Federation's main goal is to promote the wise use of natural resources through education programs, publications, and research. The NWF sponsors an annual education program called National Wildlife Week, produces science and social studies curriculums for primary and secondary schools, and publishes the annual *Conservation Directory*, which is the listing of who's who in the American conservation movement.

The NWF has 650 paid staff members and a good number of these positions are located at its national headquarters in Washington, DC. The Washington office is concerned mainly about research, fundraising, lobbying, outreach to other environmental organizations, and research activities. There are independent NWF affiliates located in every state and these organizations are concerned mainly with local grassroots activities. These independent affiliates each maintain a small staff and actively recruit volunteers, sponsor activities and outings, and offer a limited number of internships. The NWF sponsors the Resource Conservation Alliance, which is a grassroots network of members and a great source of activist information. For information on the NWF affiliate in your area call the Washington office, or consult the *Conservation Directory*.

Greenpeace is a highly successful, nonviolent, direct-action environmental organization dedicated to positive environmental change. Greenpeace, which started out over 20 years ago as a small band of concerned environmentalists protesting nuclear testing, has evolved into a huge international grassroots-based organization with offices in over 20 countries and a research station in Antarctica. Its U.S. delegation is headquartered in Washington, DC, and it has regional offices in large cities like Chicago, San Francisco, Miami, New York, Boston, and Seattle. At present, Greenpeace is focusing its attention on halting the threat of toxic pollutants; protecting endangered species like whales, dolphins, seals, and land animals; and fighting to halt the depletion of the ozone layer and global warming.

Greenpeace has over 200 staff members in the United States working at its main and regional offices. These individuals are involved primarily with fundraising and grassroots organizing. The organization strongly encourages the inclusion of people of color. In addition, Greenpeace has on its staff a group of conservation specialists, photographers, and professional outdoorsmen to carry out and publicize its campaigns. Greenpeace has a large group of paid canvassers who conduct door-to-door donation drives. There are also a large number of volunteer opportunities in the organization. Greenpeace encourages grassroots action by its supporters and provides assistance to hundreds of local causes.

The National Audubon Society (NAS) is a group of citizen and professional conservationists whose goal is to save wildlife and their habitats. While this group has in the past been concerned mainly with the protection of birds and their habitats, the NAS has broadened its concern to include the conservation of land and water, pollution issues, energy policy, and global environmental concerns. The NAS works toward these goals through activities such as lobbying, litigation, scientific research, sanctuary management, education programs, publications, and film documentaries. The Audubon Adventure Club sponsors educational programs for over 500,000 elementary school children in 14,000 classrooms annually.

The National Audubon Society has about 320 employees working at its headquarters in New York City and its nine regional offices in Anchorage, Alaska; Columbus, Ohio; Camp Hill, Pennsylvania; Albany, New York; Boulder, Colorado; Tallahassee, Florida; Austin, Texas; Manhattan, Kansas; and Sacramento, California. There are also a number of part-time and seasonal employment opportunities. In addition, the NAS offers seasonal paid internships. There is a legal intern program in Washington, DC, and field intern opportunities at Audubon sanctuaries nationwide, where interns learn valuable conservation and teaching skills. For information on internships contact the Human Resources Department in New York. There are also volunteer opportunities at 500 Audubon membership chapters and sanctuaries.

The World Wildlife Fund (WWF) is an international conservation organization dedicated to protecting endangered wildlife and wildlands. The WWF is an affiliate of the international WWF network and is one of 23 worldwide organizations. The WWF has developed a list of nine goals that it hopes to achieve: to protect individual species; to protect habitat; to influence public opinion and the policies of governments and private institutions; to support scientific investigation; to promote education in foreign countries; to offer training to local wildlife professionals; to encourage self-sufficiency in developing nations; to monitor international wildlife trade; and to promote ecologically sound development. At present, a large share of its resources are being used in Latin America where the WWF has helped create and monitor a number of national parks and wildlife reserves.

The WWF employs 300 environmental professionals and its United States office is located in Washington, DC. The Washington office is primarily involved with fundraising, research activities, and the management of field activities. A large number of its professional employees have advanced training in the conservation sciences and work at the various field stations throughout the world. The WWF maintains a job line at (202) 861-8350, with employment opportunities that are updated weekly. There are a limited number of summer internships and one-year research fellowships available. At present there are no organized volunteer activities in the United States.

The Sierra Club is one of the oldest, most respected and active environmental organizations in the United States. With 57 chapters in almost every state and 375 local groups, the Sierra Club is likely to have an office somewhere near you. At present, their main activities include working for clean air and water, designating more protected areas of wilderness in national parks and forests, supplying information on issues like global warming and acid rain, and promoting environmental education programs.

There are 328 paid staff positions in the Sierra Club. While a number of jobs are at the Sierra Club's headquarters, located in San Francisco, California, there are employment opportunities at every chapter. The San Francisco office acts as the information nerve center of the club's activities. The Sierra Club maintains a philosophy of local activism, and the chapters maintain a high level of autonomy from the national headquarters and concentrate mostly on local issues. In every chapter, volunteers play a key role in the club's activities. With over 350 volunteer groups located throughout the United States, members participate in lobbying efforts, organize campaigns, and lead letter writing efforts. In addition, the club promotes member interest in wilderness experiences and sponsors many backpacking, canoeing, skiing, and bicycling trips. Thus, staff, volunteers, and members have the opportunity to share their experiences and concerns.

The Wilderness Society is devoted to the preservation of wildlife and the wilderness, and concentrates these efforts in government protected lands like state and national forests, parks, deserts, rivers, and shorelands. The society concentrates its efforts on persuading the government to designate more public land as wilderness areas, which are off-limits to any type of development or alteration. They are also conducting an intensive legislative campaign in Congress to reduce the level of funding for road building in national forests, to reduce the amount of logging on federal lands, and to save ancient forests. In conjunction with this project is a study to track the transition of jobs in logging communities so that economic hardships in these communities can be averted.

The Wilderness Society maintains its headquarters in Washington, DC, and has 14 field offices located throughout the United States. The society has a staff of 135 environmental professionals working in five areas: membership development; finance; administration; resource and economic planning; and conservation. There is also a congressional lobbying staff in Washington and most of these individuals have previous experience working on Capitol Hill. There are volunteer and unpaid intern opportunities at each office. Inquiries should be made to the main office.

The Natural Resource Defense Council (NRDC), formed by a group of Yale Law School classmates in 1970, fights for environmental justice in the courts of law. The NRDC has described its mission and approach to environmental problems as follows: "The power of the law. The power of science. The power of the people. In defense of the environment." The NRDC is presently involved in a number of issues that include global warming, air and water pollution, energy policy, the preservation of coastal environments, and the protection of rainforests. Its staff, in addition to a corps of lawyers, includes conservation scientists, research specialists, and staff assistants. In addition, the NRDC publishes the monthly *Amicus Journal* magazine, which is a wonderful source of environmental information.

The NRDC maintains a staff of 160 at its national headquarters in New York and at four regional offices in Washington, DC; Los Angeles, California; San Francisco, California; and Honolulu, Hawaii. There is a paid summer intern legal program in New York, as well as a limited number of general internships at each office. For information on volunteer opportunities contact the director of volunteers in New York.

The Environmental Defense Fund (EDF) is an organization of lawyers, engineers, scientists, and economists, working to improve the state of the environment and to protect public health. The EDF has identified eight areas to focus its concern: toxic waste, energy, acid rain, wildlife conservation, global warming, ozone depletion, biotechnology, and water resources. In addition to these concerns, the EDF runs the Environmental Information Exchange, which is a clearinghouse of legal, economic, and scientific information. They also publish a large number of reports each year on a variety of environmental issues. Through sustained legal pressure, backed by solid expertise on environmental issues, the EDF has earned a reputation as a tough and effective voice in the environmental arena.

The EDF makes its headquarters in New York City and has five regional offices in Washington, DC; Oakland, California; Boulder, Colorado; Raleigh, North Carolina; and Austin, Texas. There are presently 145 staff members nationwide. In addition to support staff, the EDF hires a number of lawyers, conservation scientists, engineers, lobbyists, economists, publicists, and policy analysts. The EDF offers paid internships at each of its offices. Interns are typically upper-level college undergraduates or graduate students, and they work on issues such as global climate change, recycling, fundraising, and education. In some cases students can earn college credit for their work. In order to apply for an internship, send a resume, writing sample, transcript, and a letter describing why you would like to work for the EDF to its national headquarters.

Direct Action Organizations

There are other, less formal organizations that prefer the use of direct action tactics over the more traditional methods of negotiation and concession making used by most of the larger environmental organizations. These groups take polluters and those considered "environmental criminals" head-on and have used the media and public attention very effectively in achieving their goals.

Earth First! uses militant-style tactics to stop logging in the Pacific Northwest, to greatly expand the acreage of designated wilderness area, and to protect endangered species like the grizzly and northern timber wolf. The central theme of Earth First! is that there is absolutely "no compromise in the defense of Mother Earth!" They use tactics such as chaining themselves to trees to block bulldozers and clear-cutting, cutting power lines, and tree spiking to stop environmental destruction. There are 27 chapters of Earth First! throughout the country and you can get information on volunteer opportunities by writing or calling Earth First! P.O. Box 5871, Tucson, Arizona 85703, phone (602) 622-1371.

The Sea Shepherd Conservancy describes itself as a "policing body" that enforces international regulations against the illegal slaughter of seals, dolphins, and whales. This group conducts educational programs to increase public awareness of endangered sea animals and also uses such tactics as the blocking, ramming, and sinking of ships that intend harm to sea animals. The Sea Shepherd Conservancy is an all-volunteer organization and everyone from office personnel to sea captains and mates is a volunteer. For further information write or call The Sea Shepherd Conservation Society, P.O. Box 7000–S, Redondo Beach, California 90277, phone (203) 373-6967.

The Rainforest Action Network (RAN) is working to bring the destruction of the world's rainforests to the public's attention through direct actions like boycotts and information campaigns. RAN has established the Rainforest Action Groups in 150 locations around the United States and organizes local citizens for nonviolent action. RAN also coordinates an information network with over 60 environmental and human rights groups worldwide. RAN offers a number of internships at its main office and at field locations in and around the rainforests of the world. For further information contact: Rainforest Action Network, 450 Sansome Street, Suite 700, San Francisco, California 94111, phone (415) 398-4404.

Environmental Resource Centers (Regional)

The best way to conduct a job information search in your area is to contact regional and local environmental organizations. The following is a list of environmental centers located throughout the country that have information on local and regional environmental issues, jobs, internships, and volunteer opportunities.

Ecology Center
2530 San Pablo Ave.
Berkeley, CA 94702
(510) 548–2220

Alaska Center for the Environment
519 W. 8th St.
Anchorage, AK 99501
(907) 274-3621

The Ecology Center
101 East Broadway, Room 602
Missoula, MT 59802
(406) 728-5733

Ann Arbor Ecology Center
417 Detroit St.
Ann Arbor, MI 48104
(313) 761-3186

Earthright Institute
Gates-Briggs Building, Rm 322
White River Junction, VT 05001
(802) 295-7734

Environmental Action
1525 New Hampshire Avenue NW
Washington, DC 20036
(202) 745-4870

Louisiana Action Network
P.O. Box 663323
Baton Rouge, LA 70896
(504) 928-1315

Volunteer and Internship Organizations

These organizations place students and educators in internships and volunteer positions all around the country. The following two have been very successful and have built a good reputation in placing students in good environmental positions.

The Environmental Careers Organization (ECO)—formerly the CEIP Fund—is a highly respected organization that places college students (both undergraduate and graduate) with at least three years of credit, in paid intern positions. These jobs last anywhere from three months to two years, and in many cases lead to employment in the organization. The ECO estimates that 80 percent of its interns are hired after their initial job period. The majority of jobs are in private companies and government organizations. Intern positions are highly competitive and only one in eight applicants are placed in a job. There are four regional offices located in Cleveland, Ohio; San Francisco, California; Seattle, Washington; and Tampa Bay, Florida.

In addition, the ECO distributes all types of information on environmental organizations, job search strategies, and resources. They also publish *The Complete Guide to Environmental Careers,* which is a valuable book for anyone interested in an environmental career. A new edition of this book is due out in 1993. For an application and more information contact:

The Environmental Careers Organization
68 Harrison Ave.
Boston, MA 02111
(617) 426-4375

The Student Conservation Association (SCA) offers internships to high school and college students, teachers, senior citizens, and anyone else interested in helping manage public lands in the United States. The SCA has two management programs: the High School Work Group and the Resource Assistance Program. When writing for information please specify which program you are interested in. This group is also active on many college campuses.

The SCA also publishes *Earth Work,* a monthly listing of environmental and natural resource management jobs, with information on internships and volunteer positions. For more information contact:

The Student Conservation Association
P.O. Box 550
Charleston, NH 03603

Additional Sources of Information

Directory of Natural Science Centers (1990). National Science for Youth Foundation, 130 Azalea Dr., Roswell, GA 30075, (800) 992–6793 or (404) 594-9307. Gives detailed information on over 1,350 nature centers throughout the country.

New Careers: A Directory of Jobs and Internships in Technology and Society. Student Pugwash USA, 1638 R. St. NW, Suite 32, Washington, DC 20009, (202) 328-6595. Published every even numbered year. $18 nonstudent, $10 student annually. Full details on when and how to apply for internships and entry-level jobs with nonprofit organizations. Some of the types of organizations covered are environment and energy, development, communication, peace and security, health, law, and general science.

Summer Employment Directory of the United States (annual). By Peterson's Guides, PO Box 2123 Princeton, NJ 08543-2123. $14.95. Over 75,000 summer jobs at resorts, camps, national parks, and government offices, many with an environmental bent.

Great Careers: The Fourth of July Guide to Careers, Internships, and Volunteer Opportunities in the Nonprofit Sector (1990). By Devon Smith and James LaVeck and published by Garrett Park Press.

The National Directory of Internships (1991). By Timothy Stanton and Kamil Ali and published by The National Society for Internships and Experiential Education. This is a guide that describes how to find an intern program that is right for you, and how to develop that experience into a career.

Internships: The Guide to on the Job Training Opportunities for Students and Adults (1992). Published by Peterson's Guides. Describes strategies for finding information on internships and helpful tricks to use for the application process. Included is a chapter on environmental organizations with descriptions of over 90 organizations around the United States that offer internships.

Volunteer USA (1991). By Andrew Carroll and published by Fawcett Columbine. A good general directory of volunteer organizations in the United States. Included is a chapter on environmental volunteer opportunities and a listing of organizations.

Volunteer! The Comprehensive Guide to Voluntary Services in the U.S. and Abroad. By the Council on International Exchange Services, Campus Services, 205 E. 42nd St., New York, NY 10017.

Helping Out in the Outdoors: A Directory of Volunteer Jobs in State and National Forests. Northwest Trails Association, 16812 36th Ave. West, Lynwood, WA 90836.

Conservation Directory (annual). By the National Wildlife Federation. Includes a state by state listing of nonprofit volunteer organizations, with full descriptions and contact numbers, addresses, and names.

Environmental Magazines

Magazines are a great source of environmental job information because they list thousands of jobs and intern opportunities every year. Many of the following are monthly or quarterly publications that specialize in environmental issues. Some of these magazines are available at local newsstands, and most can be found at community and college libraries or at local environmental organizations.

Buzzworm
   P.O. Box 6853
   Syracuse, NY 13217-7930
   An independent environmental magazine with a 4–8 page monthly listing of internships, fellowships, and jobs.
Community Jobs
   ACCESS
   50 Beacon St.,
   Boston, MA 02108
   (617) 720-2627
   This is one of the best and most inclusive listings of nonprofit jobs nationwide.
EarthWork
   (Formerly JobScan)
   EW Dept. 517402
   Student Conservation Association
   P.O. Box 550
   Charlestown, NH 03603-0550
   (603) 826-4301
   A monthly magazine stuffed with information on full-time and seasonal jobs in administration/management, education, field work, policy, and research. Subscriptions $29.95.
Environmental Job Opportunities
   550 N. Park Street
   15 Science Hall
   University of Wisconsin-Madison
   Madison, WI 53706
   (608) 263-1815
Environmental Careers
   PH Publishing Inc.
   760 Whalers Way, Suite 100A
   Fort Collins, CO 80525
   (303) 229-0029
   Subscription $28.00.

Environmental Opportunities
    P.O. Box 969
    Stowe, VT 05672
    (802) 253-9336

International Employment Hotline
    P.O. Box 3030C
    Oakton, VA 22124
    Subscription $36.00.

Jobs Available
    P.O. Box 1040
    Modesto, CA 95353

Jobs Clearinghouse
    The Association for Experiential Education
    CU Box 249
    Boulder, CO 80309
    Publishes a monthly listing of intern and outdoor education teaching opportunities with an emphasis on wilderness experience.

The Job Seeker
    Route 2, Box 16
    Warrens, WI 54666
    Subscription $60 per year. Published biweekly and lists over 200 job openings.

Western Environmental Job Letter
    Box 660
    Wellington, CO 80549
    Subscription $24.00.

Opportunities
    Natural Science for Youth Foundation
    130 Azalea Dr.
    Roswell, GA 30075
    (800) 992-6793 or (404) 594-9367
    $35 annually. 45–70 environmental jobs listed per issue primarily at nature centers throughout the United States.

Caretaker Gazette
    P.O. Box 342
    Carpentersville, IL 60110
    (708) 658-6554
    Published quarterly. Around 60 job announcements and internships.

Environmental Career Bulletin
    11693 San Vicente Boulevard, Suite 327
    Los Angeles, CA 90047
    (213) 399-3533
    Free. 150–200 jobs advertised per issue, mostly jobs in the private sector and some nonprofit organizations. Also lists upcoming environmental job fairs nationwide.

The Wildlifer
Wildlife Society
5410 Grosvenor Lane
Bethesda, MD 20814
(301) 897-9770
Bimonthly, $33 annually.

Environmental Computer Bulletin Boards

Computers have infiltrated our lives in countless ways, and the environmental field is no exception. The following are companies that make their data on environmental issues, including job openings, available to home computer users. In general, these companies charge anywhere from $5 to $25 per hour of connection time. For a detailed description on using the electronic media as an aid in your job search, read *Ecolinking: Everyone's Guide to Online Environmental Information* (1992), by Don Rittner and published by Peachtree Press. The following are some of the most popular computer information organizations that deal specifically with environmental issues.

EcoNet
18 Deboom St.
San Francisco, CA 94107
(415) 442-0220

EarthNet
P.O. Box 330072
Kahului, Maui, HI 96733
(808) 872-6090

EnviroNet
Building E.
Fort Mason, CA 94123
(415) 474-6767
Sponsored by Greenpeace Action.

The Well
27 Gate Rd.
Sausalito, CA 94965
(415) 332-4335

Environmental Search
Services and Job Lines

JOBsource. Colorado State University, College of Natural Resources, 418 South Howes St., Suite D, Fort Collins, CO 80523 (1-800-727-JOBS). This organization uses computer data searches to match applicants with environmental jobs. There is a $30 fee per search and they guarantee 6 to 25 job matches per search.

Dial an Internship/Job. National Association of Interpretation. Tape recorded advertisements for full-time, seasonal, and temporary employment and internships for environmental jobs. For jobs dial (303) 491-6434, for internships dial (303) 491-6784.

The Association of Interpretive Naturalists Telephone Hotline. (301) 948-8844. Lists jobs and internships for outdoor educators. Openings are listed alphabetically by state.

Additional Sources

Moore, Roger L.; LaFarge, Vicki; and Tracy, Charles L. *Organizing Outdoor Volunteers.* Boston: AMC Books, 1992.

Stienstra, Tom. *Careers in the Outdoors.* New York: Foghorn Press, 1992.

Cowan, Jessica, ed. *A Guide to Careers in Social Change.* New York: Barricade Books, 1991. A listing of national social change organizations, mostly nonprofit, with a healthy listing of environmental organizations.

McAdam, Terry W. *Doing Well by Doing Good: The Complete Guide to Careers in the Nonprofit Sector.* Rockville, MD: Fund Raising Institute, 1991.

SEVEN

Green Communication

Communicating about the Environment

If you are reading this book, chances are that you consider yourself to be an environmentalist. It could be that you are just beginning the process of educating yourself about the pressing environmental issues at hand or that you know a lot about the environment and are getting ready to make saving our earth your life's work. No matter what your situation, a large part of your environmental education has come from the media and various other communication sources, and not from teachers or national parks.

In almost every major newspaper, there is usually a piece on an environmental issue. Businesses publish literature on the "naturalness" and environmental policies of their organizations, and television advertising sends messages to us about saving our planet. The people who work to continually educate us are key components of environmental education in this country. They include journalists for newspapers, magazines, organizational publications, free-lance writers, technical writers, and community relations personnel. These people work at all levels of government, corporations, and consulting firms, and the need for them is growing. In addition, an increasing number of positions are available in electronic media such as radio, television, film, and video. In all of these areas, there are numerous other people working to support the green communicator. Researchers, writers, and production staffs all play an important role in spreading the word.

There is an extraordinary amount of information about the environment that needs to be communicated. Each day there are new catastrophes, new findings, new developments, and new policies that need to be presented to the general public. In many cases the transmission of information is mandated by law in the environmental field. For example, when a major polluter is ordered to clean up its facilities or dump sites, it is required by law to share with the public every detail of its plan for how it is going to go about doing its work. Community right-to-know legislation gives the individual the opportunity to evaluate the risks and hazards associated with any project.

Preparing for a Career as an Environmental Communicator

In preparing for a career as an environmental communicator, you will need to consider the many factors that you would be dealing with on an everyday basis such as: an overwhelming amount of information, sensitive issues, resistance to change, and a growing but relatively new environmental awareness on the part of industrial and corporate America.

Individuals who work in communications generally do not have a technical background. More typically, these individuals have been trained to understand the information needs of the general public or of employees. No matter what the profession, the work demands a willingness to learn continually and an ability to convey oral and written information in an understandable manner. It also requires keeping up on environmental legislation, public opinion about the environment, and new developments in environmentally related areas such as research and development, violations, policies, and trends.

A general Bachelor of Arts degree in English, sociology, or geography coupled with an interest in the environment is one way to prepare yourself. Others may seek more specialized degrees in environmental studies, communications, journalism, technical writing, or marketing. Whatever course of study you pursue, an essential part of your education will be your independent study of environmental issues.

Salaries, while generally and traditionally lower than those in highly technical professions, are reasonable. Starting salaries can range from anywhere in the mid teens to the upper thirties depending on the size of the organization and the level of your experience.

Opportunities in Environmental Communication

The following list of jobs in the communication industry illustrates where opportunities exist and where new opportunities are developing for the career-minded environmentalist. As you will note, some of these careers have not traditionally been considered to be green careers, but your interest and commitment to the environment can make them that way.

Written Communication: News

As you begin your investigation to find an environmental career, you will find yourself buying, reading, saving, and clipping literature from a variety of sources. This information while educational and informative is also representative of a possible career. Under the umbrella title of "journalism" exist many opportunities for good writers.

Environmental Reporter

Newspapers, in addition to being an excellent source for names, organizations, agencies, corporations, and individuals to contact in your career planning and job search, also are the source of a new and expanding career. The environmental reporter is some-one who focuses exclusively on writing about the environment. It needs to be noted however, that presently, this is not where the most jobs are. The *New York Times*, a newspaper with one of the largest circulations in the country, has only one environmental reporter on staff. Smaller local and community newspapers may cover environmental issues but the staff reporters cover many areas and do not focus exclusively on the environment.

Environmental Writers:
Magazines, Journals, Newsletters

In response to increased concern and desire for information about the state of our earth, numerous new publications have come into existence that focus exclusively on the environment. Some, like *Biocycle, The Journal of Waste Recycling,* focus on a specific issue and are very specialized. Others, like *E: The Environmental Magazine,* are designed to convey information and news about a wide variety of environmental issues, controversies, and general interest stories.

Most of these publications hire full-time staff writers to pro-duce stories for each issue whether it is weekly, monthly, or bimonthly, and numerous free-lance writers are called upon to cover and report on special interest groups and issues.

Whether free-lance or full-time, the opportunities for envi-ronmental writers are continuing to grow. The following list of publications and their addresses will give you a general idea of the number, type, and variety of publications that exist today. This list is by no means comprehensive but instead is a sampling of some of the best magazines, journals, newsletters, and organ-izational publications that are available. Many of these are avail-able at newsstands, others in libraries, and others directly from the organizations themselves. Addresses have been included for those who are interested in requesting further information.

Environmental Publications

Amicus Journal
40 West 20th Street
New York, NY 10011
(212) 727-2700

Air & Waste Management Association
P.O. Box 2861
Pittsburgh, PA 15230
(412) 232-3444

Biocycle, The Journal of Waste Recycling
The JG Press
P.O. Box 351
18 South Seventh Street
Emmaus, PA 18049
(215) 967-4135

Buzzworm, The Environmental Journal
2305 Canyon Blvd., Suite 206
Boulder, CO 80302
(303) 442-1969

Clean Yield Publications
Box 1880
Greensboro Bend, VT 05842
(802) 533-7178

E: The Environmental Magazine
Earth Action Network
28 Knight Street
Norwalk, CT 06851
(203) 825-0061

Earth First! The Radical Environmental Journal
P.O. Box 5871
Tucson, AZ 85703

Earthwork Magazine
Student Conservation Association
P.O. Box 550
Charlestown, NH 03603
(603) 826-4301

ECOSOURCE
Box 1270
Guelph, Ontario, Canada
N1H 6N6
(519) 763-8888

Environmental Business Journal
Environmental Business Publications, Inc.
827 Washington
San Diego, CA 92103
(619) 295-7685

Garbage: The Practical Journal for the Environment
Old House Journal Corp.
Glouster, MA 01930
(508) 283-3200

The Green Business Letter
Tilden Press Inc.
1526 Connecticut Avenue, NW
Washington, DC 20036
(800) 955-GREEN

Hazardous Waste Material
Hazardous Waste Control Institute
7237 Hanover Pky.
Greenbelt, MD 20770
(301) 982-9500

In Business, The Magazine for Environmental Entrepreneuring
The JG Press Inc.
P.O. Box 323
18 South Seventh Street
Emmaus, PA 18049
(215) 967-4136

Journal of Environmental Health
National Environmental Health Association
720 South Colorado Blvd.
Denver, CO 80222
(303) 756-9090

National Parks
National Parks and Conservation Association
1015 31st Street, NW
Washington, DC, 20007
(202) 944-8530

Organic Times
New Hope Communication
1301 Spruce Street
Boulder, CO 80302

Pulp & Paper
500 Howard Street
San Francisco, CA 94105
(415) 397-1881

Rachael's Hazardous Waste News
Environmental Research Foundation
P.O. Box 3541
Princeton, NJ 08543

Research Reports
The Council on Economic Priorities
30 Irving Place
New York, NY 10003

Science
American Association for the
  Advancement of Science
1333 H Street NW
Washington, DC 20005
(202) 326-6400

Wastelines
Environmental Action Foundation
1525 New Hampshire Avenue NW
Washington, DC 20036
(301) 891-1100

Whole Earth Catalogue
Whole Earth Access
2990 Seventh Avenue
Berkeley, CA 94710
(800) 845-2000 or (415) 845-3000

Written Communication: Books

A new and growing market for environmental literature has
opened up as a result of our increased awareness about the state
of the earth. From children's books to reference books to college
textbooks, literature about the environment is in demand. The
writers of these books come from a multitude of backgrounds, but
all the writers have one thing in common—a desire to educate
and inform the masses about the problems we face and to pose
possible solutions.

The skills required to write books are similar to those needed
for any communication career: an interest in the subject, an
ability to translate an overwhelming amount of information into
a manageable and understandable format, and an ability to meet
deadlines. Writing a book also requires that the author have a
good sense of what types of literature exist already and a good
sense of what people are interested in reading. The following
three categories are areas that are publishing more and more about
the environment.

Children's Books

Children's book publishing is a very large segment of the publishing industry in America. There are a number of large publishing houses that have children's divisions as well as a number of small, independent presses. Manuscripts for children's books are accepted over the transom, which means that manuscripts or ideas can be sent directly to the publishers without having to hire a literary agent.

If you are interested in writing children's books you will need to spend some time browsing in bookstores to see what type of literature is being published. You will also want to notice what types of books are coming from what houses and send your manuscript to the publishers that are already publishing environmental literature. The best directory for getting addresses is the *Literary Marketplace* (available in the reference section of most libraries).

Trade and Reference Books

A new market has emerged for trade and reference titles that concern themselves with environmental issues—from books on specific environmental issues to books on careers for environmentalists and everything in between. Many of these books are written by experts, by academics, by professionals, and sometimes in conjunction with specific organizations.

Three of the best publishers of environmental literature are as follows:

Island Press
Box 7
Covelo, CA 95428
(800) 828-1302

Island Press, founded in 1978, is a nonprofit organization that publishes books solely on environmental topics and publishes 20 to 30 titles a year. Its backlist of 87 title includes such classics as Herman Daly's *Steady State Economics* and *Rush to Burn: Solving America's Garbage Crisis?* from Newsday.

Natural Resources Defense Council
40 West 20th Street
New York, NY 10011

The Natural Resources Defense Council publishes books and papers on environmental issues such as the rainforest, solid waste management, water and air pollution, energy issues, food safety, pesticide use, and experimental education material for children. The NRDC is also an excellent resource for those interested in researching a specific environmental issue.

World Resources Institute
1750 New York Avenue, NW
Washington, DC 20006

The WRI does research and assists the government, private sector, and organizations on environmental and management issues. The *World Resource Report* is published by the institute and will be a useful source of information for any type of literature that you are considering writing.

Having a good idea coupled with previous writing experience is the beginning of a career as a writer. In many instances it may be a good idea to work with an agent to find the best home for your project.

College Textbooks

Many disciplines in the college curriculum have courses that deal specifically with environmental issues. In economics departments there are courses in environmental and natural resource economics, and at many institutions an introductory ecology course is offered. In addition, many English instructors are using collections of environmental literature to teach their writing courses. To write a college textbook means that the writer has to have a master's degree or Ph.D. in the discipline to be a qualified author. Like children's books, manuscripts for college textbooks are accepted over the transom. The *Literary Marketplace* (LMP) is the best resource for the addresses of publishers.

Written Communication: Technical Writer

The technical writer, or proposal writer, is the title given to someone who has the responsibility of gathering together technical information from a variety of sources in preparation for making a bid on a government contract. The better written the document, the better the chance of getting the contract. Due to an increase in environmental legislation, in addition to presenting a polished document, it is also essential that it is environmentally focused.

The skills required for this type of work include the ability to convert records and oral reports into finished documents, an eye for detail, and often specialized training in technical writing. Many colleges now offer degrees in technical writing, or courses in technical writing are often offered through English departments.

Electronic Communication: Television, Video, Film

The electronic media of television, video, and film are all an integral part of conveying environmental messages to the general public. From consultants who are hired to guarantee and protect the locations for shoots to writers and producers of documentaries about nature and the environment, the electronic media are filled with environmentally literate workers. Training for some of these careers is extensive and technical, and the environmental component is often the result of a strong personal interest and desire to work for the benefit of the earth. Consultants are used to provide the expertise to confirm and polish the messages that will eventually reach a large viewing audience.

Consumer Communication: Advertising

More and more products are hitting the shelves with the green consumer in mind, and advertising is used to let us know about it. Companies that have responded to our pleas for safer products are gaining marketshare. As a result, advertising agencies are called upon to promote this new consumer greenness. While a few have taken advantage of this consumer trend by promoting products as green when in fact they are not, there does exist a real need for meaningful, effective advertising of environmentally safe products and services. Designers, account executives, and copywriters are all needed to develop and design environmental advertising campaigns.

Goodwill Communication: Community Relations Manager

Community right-to-know laws require manufacturers to reveal detailed information on the expected risks of a clean-up action and on preparing for emergencies near chemical plants, utilities, and other factories. This information is passed to the public through a community relations manager. This is essentially a function of the public relations department. The job requirements include an ability to communicate and diplomatic and technical skills. An ability to understand scientific and technical data is also a requirement. Most community relation managers work for consulting firms that serve the offending polluters.

The Critical Link: Market Research

Whenever there is a great deal of change and concern occurring rapidly and in which large amounts of money are at stake, jobs are created for market researchers who provide the data that businesses need to make decisions. Many businesses are interested in increasing their marketshares by producing green goods and in gaining the edge in an evolving market. Market researchers are continually probing the consumer to find out just how environmentally committed we are. Because of underdeveloped technologies, recycled or green products tend to cost more. The most often asked question by researchers is whether consumers are willing to pay a premium for environmental improvement. Individuals who work in market research typically have studied market research and have taken courses in statistics.

Appendix A

General Bibliography

Anzalone, Joan. *Good Works: A Guide to Careers in Social Change.* New York: Dembner Books, 1985.

Basta, Nicholas. *The Environmental Career Guide.* New York: John Wiley and Sons, 1991.

Bolles, Richard Nelson. *What Color is Your Parachute?* Berkeley: Ten Speed Press, 1991.

Bowker, R. R. *Environmental Abstracts.* New York: Bowker A&I Publishing, 1991.

Careers in the Renewable Energy and Conservation Professions and Trades. Silver Spring, MD: U.S. Department of Energy, Conservation and Renewable Energy Inquiry and Referral Service, 1985.

Chapman, Robert B., and Johnson, Miriam. *Work in the New Economy: Careers and Job Seeking into the 21st Century.* Indianapolis, IN: JIST Works, 1989.

DeAngelis, L., ed. *The Complete Guide to Environmental Careers.* Washington, DC: Island Press, 1989.

Didsbury, Howard F. *The World of Work: Careers and the Future.* Bethesda, MD: World Future Society, 1983.

Environmental Career Organization. *Beyond the Green: Redefining and Diversifying the Environmental Movement.* Boston: The Environmental Careers Organization, 1992.

Fanning, Odom. *Opportunities in Environmental Careers.* Lincolnwood, IL: VGM Career Horizons, 1992.

Feingold, S. Norman, and Miller, Norma. *Emerging Careers: New Occupations for the Year 2000 and Beyond.* Garrett Park, MD: Garrett Park Press, 1983.

Goldberg, Dick, and Kazan, Katherine. *Careers without Reschooling: The Survival Guide to the Job Hunt for Liberal Arts Graduates.* New York: Continuum, 1985.

Hawes, Gene R., and Brownstone, Douglass L. *The Outdoor Careers Guide.* New York: Facts on File, 1986.

Hopke, William E. *The Encyclopedia of Careers and Vocational Guidance.* Chicago: J.G. Ferguson Publishing Co., 1990.

Krannich, Caryl Rae. *Almanac of International Jobs and Careers.* Woodbridge, VA: Impact Publications, 1991.

Krannich, Ronald L., and Krannich, Caryl Rae. *The Complete Guide to International Jobs and Careers.* Woodbridge, VA: Impact Publications, 1990.

Lanier-Graham, Susan. *The Nature Directory.* New York: Walker and Company, 1991.

Lewis, Adele Beatrice, and Kuller, Doris. *Fast-Track Careers for the 90's.* Glenview, IL: Scott, Foresman, 1990.

Marek, Rosanne J. *Opportunities in Social Science Careers.* Lincolnwood, IL: VGM Career Horizons, 1990.

McAdam, Terry W. *Doing Well by Doing Good: The Complete Guide to Careers in the Nonprofit Sector.* Rockville, MD: Fund Raising Institute, 1991.

Miller, Louise. *Careers for Nature Lovers and Other Outdoor Types.* Lincolnwood, IL: VGM Career Horizons, 1992.

Pira's International Environmental Information Sources. New York: Pira Information Services, 1990.

Sacharov, Al. *Offbeat Careers: The Directory of Unusual Work.* Berkeley: Ten Speed Press, 1988.

Spenner, Kenneth I.; Otto, Luther B.; and Vaughn, R. A. Call. *Career Lines and Careers.* Lexington, MA: Lexington Books, 1982.

Vacca, Susan M. *The Harvard Guide to Careers.* Cambridge, MA: Harvard University Press, 1991.

VGM's Careers Encyclopedia. Lincolnwood, IL: VGM Career Horizons, 1991.

VGM Handbook of Scientific and Technical Careers. Lincolnwood, IL: VGM Career Horizons, 1990.

Warner, David J. *Environmental Careers.* Boca Raton, FL: Lewis Publishers Inc, 1991.

Weinstein, Bob. *How to Switch Careers.* New York: Simon & Schuster, 1985.

Woodburn, John H. *Opportunities in Energy Careers.* Lincolnwood, IL: VGM Career Horizons, 1985.

Appendix B

State Agencies

ALABAMA

Conservation and Natural
  Resources Department
64 N. Union St.
Room 702
Montgomery, AL 36130
(205) 242-3486

ALASKA

Environmental Conservation
  Department
3220 Hospital Dr.
P.O. Box O
Juneau, AK 99811-1800
(907) 465-2606

ARIZONA

Environmental Quality
  Department
2005 N. Central Avenue
Phoenix, AZ 85004
(602) 257-2300

ARKANSAS

Health Department
4815 W. Markham
Little Rock, AR 72201
(501) 661-2111

Pollution Control and Ecology
  Department
8001 National Dr.
Little Rock, AR 72219
(501) 562-7444

CALIFORNIA

Environmental Affairs Agency
1102 Q St.
P.O. Box 2815
Sacramento, CA 95812

COLORADO

Natural Resources Department
1313 Sherman St.
Room 718
Denver, CO 80203
(303) 866-3311

CONNECTICUT

Environmental Protection
  Department
165 Capitol Ave.
Hartford, CT 06006
(203) 566-5599

DELAWARE

Natural Resources and
  Environmental Control
  Department
89 Kings Highway
P.O. Box 1401
Dover, DE 19903
(302) 736-4506

DISTRICT OF COLUMBIA

Public Works Department
2000 14th St., NW, 6th Floor
Washington, DC 20009
(202) 939-8000

FLORIDA

Environmental Regulation
  Department
2600 Blair Stone Rd.
Tallahassee, FL 32399-2400
(904) 488-9334

GEORGIA

Natural Resources Department
205 Butler Street, SE,
  Suite 1252
Atlanta, GA 30328
(404) 656-3530

HAWAII

Hawaiian Home Lands
  Department
P.O. Box 18789
Honolulu, HI 96805
(808) 548-6450

Environmental Health
  Administration
1250 Punchbowl St.
Honolulu, HI 96813
(808) 548-6455

IDAHO

Fish and Game Department
600 S. Walnut
P.O. Box 25
Boise, ID 83707
(208) 334-3782

ILLINOIS

Environmental Protection
  Agency
P.O. Box 19276
Springfield, IL 62794
(217) 782-3397

INDIANA

Environmental Management
  Department
105 S. Meridian St.
P.O. Box 6015
Indianapolis, IN 46206-6015
(317) 232-8162

IOWA

Natural Resources Department
Wallace Building
Des Moines, IA 50319-0034
(515) 281-5385

KANSAS

Health and Environmental
  Department
Forbes Field, Building 740
Topeka, KS 66620
(913) 296-1500

KENTUCKY

Natural Resources and
  Environmental Protection
  Cabinet
Capitol Plaza Tower, 5th Floor
Frankfort, KY 40601
(502) 564-2043

LOUISIANA

Environmental Quality
  Department
P.O. Box 44066
Baton Rouge, LA 70804
(504) 342-1222

MAINE

Environmental Protection
  Department State House
Station 17
Augusta, ME 04333
(207) 289-7688

MARYLAND

Natural Resources Department
Tawes State Office Building
Annapolis, MD 21401
(301) 974-3041

MASSACHUSETTS

Environmental Affairs
  Executive Office
100 Cambridge St., Room 2000
Boston, MA 02202
(617) 727-9800

MICHIGAN

Natural Resources Department
P.O. Box 30028
Lansing, MI 48909
(517) 373-1220

MINNESOTA

Pollution Control Agency
520 Lafayette Rd.
St. Paul, MN 55155
(612) 296-6300

MISSISSIPPI

Environmental Quality
  Department
P.O. Box 20305
Jackson, MS 39289-1305
(601) 961-5099

MISSOURI

Conservation Department
2901 W. Truman Blvd.
P.O. Box 180
Jefferson City, MO 65102-0180
(314) 751-4115

Natural Resources Department
P.O. Box 176
Jefferson City, MO 65102
(314) 751-3443

Public Safety Department
P.O. Box 749
Jefferson City, MO 65102
(314) 751-4905

MONTANA

Natural Resources and
  Conservation Department
1520 E. 6th Ave.
Helena, MT 59620-2301
(406) 444-6873

NEBRASKA

Environmental Control
  Department
State Office Building
P.O. Box 98922
Lincoln, NE 68509-8922
(402) 471-2186

NEVADA

Conservation and Natural
  Resources Department
201 S. Fall St.
Carson City, NV 89710
(702) 687-4360

NEW HAMPSHIRE

Environmental Services
  Department
6 Hazen Dr.
Concord, NH 03301
(603) 271-3503

NEW JERSEY

Environmental Protection
  Department
401 E. State St.
CN 402
Trenton, NJ 08625-0402
(609) 292-3131

NEW MEXICO

Health and Environment
  Department
1190 St. Francis Dr.
Santa Fe, NM 87503
(505) 827-0020

NEW YORK

Environmental Conservation
  Department
50 Wolf Rd.
Albany, NY 12233
(518) 457-3446

NORTH CAROLINA

Environment, Health and
  Natural Resources
  Department
P.O. Box 72687
Raleigh, NC 27611
(919) 733-4984

NORTH DAKOTA

Game and Fish Department
100 N. Bismarck Expressway
Bismarck, ND 58501
(701) 221-6300

Health and Consolidated
  Laboratories Department
600 E. Boulevard Ave.
Bismarck, ND 58505
(701) 224-2370

Parks and Recreation
  Department
1424 W. Century Ave.,
  Suite 202
Bismarck, ND 58501
(701) 224-4887

OHIO

Environmental Protection
  Agency
1800 Watermark
P.O. Box 1049
Columbus, OH 43266-0149
(614) 644-3020

OKLAHOMA

Health Department
1000 NE 10th St.
P.O. Box 53551
Oklahoma City, OK 73152
(405) 271-4200

Public Safety Department
3600 Martin Luther King Ave.
P.O. Box 11415
Oklahoma City, OK 73136-0415

Wildlife Conservation
  Department
P.O. Box 53465
Oklahoma City, OK 73152

OREGON

Fish and Wildlife Department
P.O. Box 59
Portland, OR 97207
(503) 976-6339

PENNSYLVANIA

Environmental Resources
  Department
P.O. Box 2063
Harrisburg, PA 17120
(717) 783-2300

RHODE ISLAND

Environmental Management
  Department
9 Hayes St.
Providence, RI 02908
(401) 277-6800

SOUTH CAROLINA

Health Environmental Control
  Department
2600 Bull St.
Columbia, SC 29201
(803) 734-4880

SOUTH DAKOTA

Water and Natural Resources
  Department
Joe Foss Building
523 E. Capitol
Pierre, SD 57501
(605) 773-3151

TENNESSEE

Health and Environment
  Department
344 Cordell Hull Building
Nashville, TN 37219-5402
(615) 741-3111

TEXAS

Air Control Board
6330 Highway 2901 E.
Austin, TX 78723
(512) 451-5711

Health Department
1100 W. 49th St.
Austin, TX 78756
(512) 458-7244

Parks and Wildlife Department
4200 Smith School Rd.
Austin, TX 78744
(512) 389-4800

Soil and Water Conservation
  Board
P.O. Box 658
Temple, TX 76503
(817) 773-2250

Water Commission
P.O. Box 13087
Capitol Station
Austin, TX 78711
(512) 463-5538

Water Development Board
P.O. Box 13231
Austin, TX 78711-3231
(512) 463-7847

UTAH

Health Department
P.O. Box 16700
Salt Lake City, UT 84116-0700
(801) 538-6101

VERMONT

Natural Resources Agency
103 S. Main St.
Waterbury, VT 05676
(802) 244-6916

VIRGINIA

Natural Resources Secretariat
9th St. Office Building,
  Room 525
Richmond, VA 23219
(804) 786-0044

WASHINGTON

Ecology Department
MS PV-11
Olympia, WA 98504
(206) 459-6000

WEST VIRGINIA

Air Pollution Control
  Commission
1558 Washington St., E.
Charleston, WV 25311-2599
(304) 348-4022

Natural Resources Department
1900 Kanawha Boulevard E.
Charleston, WV 25305
(304) 348-2754

WISCONSIN

Natural Resources Department
P.O. Box 7921
Madison, WI 53707
(608) 266-2621

WYOMING

Environmental Quality
   Department
Herschler Building
122 W. 25th St., 4th Floor
Cheyenne, WY 82002
(307) 777-7937

PUERTO RICO

Environmental Quality Board
P.O. Box 11488
Santurce, PR 00910
(809) 725-5140

Appendix C

State Pollution Control Offices

Air Pollution

Alabama

Env. Management
Air Program
1751 Congressman W.L.
 Dickinson Dr.
Montgomery, AL 36130
(205) 271-7700

Alaska

Dept. of Env. Conservation
P.O. Box O
Juneau, AK 99811-1800
(907) 465-2666

Arizona

Dept. of Env. Quality
Office of Air Quality
2005 N. Central
Phoenix, AZ 85004
(602) 257-2308

Arkansas

Air Division
Dept. of Pollution Control &
 Ecology
8001 National Dr.
Little Rock, AR 72209
(501) 562-7444

California

Air Resources Board
P.O. Box 2815
Sacramento, CA 95812
(916) 445-4383

Colorado

Air Pollution Control Div.
Dept. of Health
4210 E. 11th Ave.
Denver, CO 80220
(303) 331-8500

Connecticut

Air Management Unit
Dept. of Env. Protection
165 Capital Ave.
Hartford, CT 06106
(203) 566-4030

District of Columbia

Env. Admin. Regulation
Room 203
2100 Martin Luther King
    Ave. S.E.
Washington, DC 20020
(202) 404-1120

Delaware

Air Resources Section
Div. of Waste Management
State Env. Control
P.O. Box 1401
Richardson & Robbins Bldg.
Dover, DE 19903
(302) 739-4791

Florida

Air Resources Management
Div. of Env. Programs
Dept. of Env. Regulation
Twin Towers Office Bldg.
2600 Blair Stone Rd.
Tallahassee, FL 32399-2400
(904) 488-1344

Georgia

Air Protection Branch
Env. Protection Div.
Dept. of Natural Resources
205 Butler St., S.E.
Atlanta, GA 30334-2400
(404) 656-6900

Hawaii

Env. Management Div.
Dept. of Health
Clean Air Branch
P.O. Box 3378
Honolulu, HI 96801
(808) 586-4200

Idaho

Dept. of Health & Welfare
Div. of Env. Quality
Planning and Evaluation
2nd Floor
1410 N. Hilton St.
Boise, ID 83706
(208) 334-5879

Dept. of Health & Welfare,
    Permits & Enforcement
Div. of Env. Quality
3rd Floor
1410 N. Hilton St.
Boise, ID 83706
(208) 334-5898

Illinois

Div. of Air Pollution Control
Env. Protection Agency
1340 N. 9th St.
Springfield, IL 62702
(217) 782-7326

Indiana

Dept. of Env. Management
Office of Air Management
P.O. Box 6015
Indianapolis, IN 46206-6015
(317) 232-8384

Iowa

Dept. of Natural Resources
Henry A. Wallace Bldg.
900 E. Grand Ave.
Des Moines, IA 50319-0034
(515) 281-8693

Kansas

Dept. of Health and Env.
Bureau of Air Quality
Bldg. 740-Forbes Field
Topeka, KS 66620
(913) 296-1570

Kentucky

Div. of Air Pollution Control
Dept. of Natural Resources &
  Env. Protection
316 St. Clair Mall
Frankfort, KY 40601
(502) 564-3382

Louisiana

Office of Air Quality &
  Radiation Protection
Dept. of Env. Quality
P.O. Box 82135
Baton Rouge, LA 70884-2135
(504) 765-0219

Maine

Bureau of Air Quality Control
Dept. of Env. Protection
State House, Sta. 17
Augusta, ME 04333
(207) 289-2437

Maryland

Dept. of Environment
Air Management Admin.
2500 Broening Hwy.
Baltimore, MD 21224
(301) 631-3255

Massachusetts

Div. of Air Quality Control
Dept. of Env. Protection
1 Winter St., 7th Floor
Boston, MA 02108
(617) 292-5630

Michigan

Air Quality Div.
Dept. of Natural Resources
P.O. Box 30028
Lansing, MI 48909
(517) 373-7023

Minnesota

Div. of Air Quality
Pollution Control Agency
520 Lafayette Rd.
St Paul, MN 55155
(612) 296-7265

Mississippi

Air Quality Div.
Office of Pollution Control
Dept. of Env. Quality
P.O. Box 10385
Jackson, MS 39289-0385
(601) 961-5171

Missouri

Air Conservation Commission
Div. of Env. Quality
Dept. of Natural Resources
P.O. Box 176
Jefferson City, MO 65102
(314) 751-4817

Montana

Air Quality Bureau
Dept. of Health & Env. Sciences
Cogswell Bldg., Rm. A116
Helena, MT 59620
(406) 444-3454

Nebraska

Air Pollution Control Div.
Dept. of Env. Control
P.O. Box 98922
State House Station
Lincoln, NE 68509
(402) 471-2186

Nevada

Div. of Env. Protection
Dept. of Conservation &
   Natural Resources
123 W. Nye Lane
Carson City, NV 89710
(702) 687-4360

New Hampshire

Dept. of Env. Services
Air Resources Div.
64 N. Main St.
Caller Box 2033
Concord, NH 03301-2033
(603) 271-1370

New Jersey

Div. of Env. Quality
Dept. of Env. Protection
401 E. State St.
CN-027
Trenton, NJ 08625
(609) 292-6704

New Mexico

Air Quality Bureau
Env. Improvement Bureau
1190 St. Francis Drive
Santa Fe, NM 87503
(505) 827-0070

New York

Div. of Air Resources
Dept. of Env. Conservation
50 Wolf Rd.
Albany, NY 12233-3250
(518) 457-7230

North Carolina

Div. of Env. Management
Dept. of Env., Health &
   Natural Resources
P.O. Box 29535
Raleigh, NC 27626-0530
(919) 733-7015

North Dakota

Div. of Env. Engineering
Dept. of Health
1200 Missouri Ave., Rm. 304
Bismarck, ND 58502
(701) 221-5188

Ohio

Office of Air Pollution Control
Env. Protection Agency
1800 Watermark Dr.
Columbus, OH 43266-0149
(614) 644-2270

Oklahoma

Air Quality Services
Env. Health Services
Dept. of Health
1000 N.E. 10th St.
Oklahoma City, OK 73117-1299
(405) 271-5220

Oregon

Air Quality Div.
Dept. of Env. Quality
811 S.W. 6th Ave.
Portland, OR 97204
(503) 229-5359

Pennsylvania

Bureau of Air Quality Control
Dept. of Env. Resources
2nd & Chestnut Sts.
Executive House
P.O. Box 2357
Harrisburg, PA 17105-2352
(717) 787-9702

Rhode Island

Div. of Air & Hazardous
    Materials
Dept. of Env. Management
291 Promenade St.
Providence, RI 02908
(401) 277-2808

South Carolina

Bureau of Air Quality Control
Dept. of Health & Env. Control
2600 Bull St.
Columbia, SC 29201
(803) 734-4750

South Dakota

Dept. of Env. & Natural
    Resources
Joe Foss Bldg., 523 E. Capital
Pierre, SD 57501-3181
(605) 773-3153

Tennessee

Div. of Air Pollution
Dept. of Env. & Conservation
701 N. Broadway
Nashville, TN 37243-1531
(615) 741-3931

Texas

Air Control Board
12124 Park 35 Circle
Austin, TX 78753
(512) 451-5711

Utah

Dept. of Env. Control
Div. of Air Quality
Div. Air Conservation
    Committee
1950 W.N. Temple
Salt Lake City, UT 84114-4820
(801) 536-4000

Vermont

Air Pollution Control Div.
Dept of Env. Conservation
103 S. Main St.
Bldg. 3 South
Waterbury, VT 05671-0402
(802) 244-8731

Virginia

Dept. of Air Pollution Control
P.O. Box 10089
Richmond, VA 23240
(804) 786-2378

Washington

Solid & Hazardous Waste
Dept. of Ecology
4224 6th Ave.
Bldg 4, Rowesix
MS/PV-11
Olympia, WA 98504-8711
(206) 459-6322

West Virginia

Air Pollution Control
  Commission
1558 Washington St. E.
Charleston, WV 25311
(304) 348-2275

Wiconsin

Bureau of Air Management
Dept. of Natural Resources
101 S. Webster St.
Madison, WI 53703
(608) 266-7718

Wyoming

Air Quality Div.
Dept. of Env. Quality
Herschler Bldg.
122 W. 25th St.
Cheyenne, WY 82002
(307) 777-7391

Land Use

Alabama

Dept of Conservation and
  Natural Resources
64 N. Union St., Rm. 702
Montgomery, AL 36130
(205) 242-3486

Alaska

Dept. of Natural Resources
400 Willoughby
Juneau, AK 99801
(907) 465-2400

California

Dept of Forestry & Fire
  Protection
P.O. Box 944246
Sacramento, CA 94244-2460
(916) 445-3976

Colorado

Land Use Commission
Div. of Local Govt.
Dept. of Local Affairs
1313 Sherman St.
Denver, CO 80203
(303) 866-2156

Connecticut

Dept. of Env. Protection
State Office Bldg.
165 Capital Ave.
Hartford, CT 06016
(203) 566-5599

Delaware

Dept. of Natural Resources &
Env. Control
Div. of Water Resources
Richardson & Robbins Bldg.
P.O. Box 1401
89 Kings Hwy.
Dover, DE 19903
(302) 739-5409

Florida

Dept. of Community Affairs
Div. of Resources Planning &
Management
Bureau of State Planning
2740 Centerview Dr.
The Rhyne Bldg.
Tallahassee, FL 32399-2100
(904) 488-4925

Georgia

Office of Planning & Budget
Div. of Physical & Economic
Development
254 Washington St. S.W.
Atlanta, GA 30334
(404) 656-3819

Hawaii

Dept. of Business & Economic
Development and Tourism
Land Use Commission
335 Merchant St., Rm. 104
Honolulu, HI 96813
(808) 587-3822

Idaho

Dept. of Fish & Game
P.O. Box 25
600 S. Walnut
Boise, ID 83707
(208) 334-3700

Illinois

Div. of Planning
Dept. of Conservation
524 S. Second St.
Springfield, IL 62701-1787
(217) 782-4543

Indiana

Dept. of Env. Management
105 S. Meridian
Indianapolis, IN 46225
(317) 232-8162

Iowa

Dept. of Natural Resources
Henry A. Wallace Bldg.
900 E. Grand
Des Moines, IA 50319
(515) 281-8690

Kansas

Dept. of Health & Env.
Bldg. 740-Forbes Field
Topeka, KS 66620
(913) 296-1570

Kentucky

Natural Resources & Env.
   Protection Cabinet
Dept. of Env. Protection
Div. of Env. Services
18 Reilly Rd.
Fort Boone Plaza
Frankfort, KY 40601
(502) 564-3035

Louisiana

Office of Planning & Budget
P.O. Box 94095
Baton Rouge, LA 70804-9095
(504) 342-7410

Maine

Dept. of Env. Protection
Bureau of Land Quality Control
State House Station 17
Augusta, ME 04333
(207) 289-2111

Maryland

Office of Planning
301 W. Preston St., Rm. 1101
Baltimore, MD 21201
(301) 225-4500

Massachusetts

Dept. of Food & Agriculture
Bureau of Land Use
142 Old Common Rd.
Lancaster, MA 01523
(508) 792-7710

Michigan

Land & Water Management
   Div.
Dept. of Natural Resources
P.O. Box 30028
Lansing, MI 48909
(517) 373-1170

Minnesota

Env. Quality Bd.
Critical Area Program
Centennial Office Bldg.
658 Cedar St., 3rd Floor
St. Paul, MN 55155
(612) 296-2603

Mississippi

State Pollution & Control
   Office
Dept. of Env. Quality
P.O. Box 10385
Jackson, MS 39289-0385
(601) 961-5171

Missouri

Dept. of Natural Resources
Div. of Env. Quality
P.O. Box 176
Jefferson City, MO 65102
(314) 751-3443

Montana

Dept. of State Lands
Div. of Forestry
2705 Spurgin Rd.
Missoula, MT 59801
(406) 542-4300

Nebraska

Dept. of Env. Control
State House Station
P.O. Box 98922
Lincoln, NE 68509-8922
(402) 471-2186

Nevada

Dept. of Conservation &
   Natural Resources
Div. of Forestry
123 W. Nye Lane, Rm. 142
Carson City, NV 89710
(702) 687-4360

New Hampshire

Office of State Planning
2¹/₂ Beacon St.
Concord, NH 03301
(603) 271-2155

New Jersey

Dept. of Env. Protection
Div. of Coastal Resources
CN 401
501 E. State St.
Trenton, NJ 08625
(609) 292-1235

New Mexico

Env. Dept.
1190 St. Francis Dr.
Santa Fe, NM 87502
(505) 827-2850

Health Dept.
1190 St. Francis Dr.
Santa Fe, NM 87502
(505) 827-2613

New York

Dept. of State
Office for Local Govt. Services
162 Washington, Ave.,
   6th Floor
Albany, NY 12231
(518) 473-3355

North Carolina

Dept. of Env. Health & Natural
   Resources
Coastal Resources Commission
P.O. Box 27687
Raleigh, NC 27611-7687
(919) 733-2293

North Dakota

North Dakota Dept. of
Transportation
Planning Div.
608 E. Boulevard Ave.
Bismarck, ND 58505
(701) 224-2513

Ohio

Dept. of Natural Resources
Div. of Wildlife
1840 Belcher Dr.
Fountain Sq., Bldg. G
Columbus, OH 43224-1329
(614) 265-6300

Oklahoma

Dept. of Pollution Control
P.O. Box 53504
Oklahoma City, OK 73152
(405) 271-4677

Oregon

Dept. of Forestry
2600 State St.
Salem, OR 97310
(503) 378-2560

Pennsylvania

Dept. of Env. Resources
P.O. Box 2063
Harrisburg, PA 17105-2063
(717) 783-2300

Rhode Island

Coastal Resources Management
Council
Stedman Govt. Center
4808 Tower Hill Rd.
Wakefield, RI 02879
(401) 277-2476

South Carolina

Land Resources Conservation
Commission
2221 Devine St., Suite 222
Columbia, SC 29205
(803) 734-9100

South Dakota

Div. of Environmental
Regulation
Dept. of Water & Natural
Resources
Joe Foss Bldg., 523 E. Capital
Pierre, SD 57501
(605) 773-3351

Tennessee

Dir. of Parks & Recreation
Dept. of Env. & Conservation
Customs House
701 Broadway
Nashville, TN 37243-0435
(615) 742-6747

Texas

Coastal Div.
General Land Office
Land Management Dept.
Stephen F. Austin Bldg.,
Rm. 730
1700 N. Congress Ave.
Austin, TX 78701
(512) 475-1467

Utah

State Health Dept.
Div. of Env. Health
288 N. 1460 W.
P.O. Box 16690
Salt Lake City, UT 84116-0690
(801) 536-6121

Vermont

Env. Board
120 State St.
Montpelier, VT 05620-3201
(802) 828-3309

Virginia

Dept. of Housing &
   Community Development
Div. of Community Dev.
205 N. 4th St.
Richmond, VA 23219
(804) 786-4966

Washington

State Parks & Recreation
   Commission
7150 Cleanwater Lane
Mail Stop KY-11
Olympia, WA 98504
(206) 753-5755

West Virginia

Office of Real Estate
   Management
Dept. of Natural Resources
Bldg. 3, Capitol Complex,
   Rm. 643
Charleston, WV 25305
(304) 348-2224

Wisconsin

Dept. of Development
123 W. Washington Ave.
P.O. Box 7970
Madison, WI 53707
(608) 266-1018

Wyoming

Dept. of Env. Quality
Land Quality Div.
Hershler Bldg.—3rd Floor West
122 W. 25th St.
Cheyenne, WY 82002
(307) 777-7756

Solid Waste

Alabama

Land Program
Dept. of Env. Management
Hazardous Waste Branch
1751 Congressman W.L.
   Dickinson Dr.
Montgomery, AL 36130
(205) 271-7700

Alaska

Div. Env. Quality
Dept. of Env. Conservation
P.O. Box O
Juneau, AK 99811-1800
(907) 465-2666

Arizona

Dept. of Env. Quality
Hazardous & Solid Waste Unit
2005 N. Central
Phoenix, AZ 85004
(602) 257-2176

Arkansas

Solid Waste Management Div.
Dept. of Pollution Control &
   Ecology
8001 National Dr.
Little Rock, AR 72209-8913
(501) 562-7444

California

Integrated Waste Management
   Board
1020 9th St., Suite 300
Sacramento, CA 95814
(916) 322-3330

Colorado

Hazardous Materials & Waste
   Management
Dept. of Health
4210 E. 11th Ave.
Denver, CO 80220
(303) 320-8333

Connecticut

Solid Waste Management Unit
Dept. of Env. Protection
18-20 Trinity St.
Hartford, CT 06106
(203) 566-5847

Delaware

Solid Waste Authority
1128 S. Bradford St.
Dover, DE 19903
(302) 739-5361

District of Columbia

Dept. of Public Works
Public Space Maintenance
   Admin.
2750 S. Capitol St., S.E.
Washington, DC 20032
(202) 727-4825

Florida

Solid Waste Management
   Program
Dept. of Env. Regulation
Twin Towers Office Bldg.
2600 Blair Stone Rd.
Tallahassee, FL 32399-2400
(904) 922-6104

Georgia

Land Protection Branch
Div. of Env. Protection
Dept. of Natural Resources
205 Butler St., SE
Floyd Towers East
Suite 1154
Atlanta, GA 30334
(404) 656-2833

Hawaii

Env. Management Div.
Clean Water Branch
P.O. Box 3378
Honolulu, HI 96801
(808) 586-4309

Idaho

Hazardous Materials Bureau
Div. of Env. Quality
Dept. of Health and Welfare
1410 N. Hilton
Boise, ID 83706
(208) 334-5879

Illinois

Div. of Land Pollution Control
Env. Protection Agency
2200 Churchill Rd.
Springfield, IL 62794
(217) 782-6760

Indiana

Dept. of Env. Management
Office of Solid & Hazardous
  Waste
105 S. Meridian, PO Box 6015
Indianapolis, IN 46206-6015
(317) 232-3210

Iowa

Div. of Air Quality & Solid
  Waste Protection
Dept. of Natural Resources
Henry A. Wallace Bldg.
900 E. Grand
Des Moines, IA 50319
(515) 281-4968

Kansas

Bureau of Air and Waste
  Management
Dept. of Health & Env.
Forbes Field-Bldg. 740
Topeka, KS 66620-0001
(913) 296-1590

Kentucky

Div. of Waste Management
Dept. of Natural Resources &
  Env. Protection
18 Reilly Rd.
Frankfort, KY 40601
(502) 564-6716

Louisiana

Dept. of Env. Quality
Solid Waste Div.
P.O. Box 82178
Baton Rouge, LA 70884-2178
(504) 765-0249

Maine

Bureau of Land Quality Control
Dept. of Env. Protection
State House Station #17
Augusta, ME 04333
(207) 289-2111

Maryland

Waste Management &
Enforcement Program
Dept. of Env.
2500 Broening Hwy.
Baltimore, MD 21224
(301) 631-3304

Massachusetts

Div. of Env. Solid Waste
Management
Dept. of Env. Management
1 Winter St.
Boston, MA 02108
(617) 292-5869

Michigan

Waste Management Div.
Dept. of Natural Resources
P.O. Box 30028
Lansing, MI 48909
(517) 373-0540

Mississippi

State Office of Pollution
Dept. of Env. Quality Control
P.O. Box 10385
Jackson, MS 39289-0385
(601) 961-5171

Missouri

Waste Management Program
Div. of Env. Quality
Dept. of Natural Resources
P.O. Box 176
Jefferson City, MO 65102
(314) 751-3176

Montana

Dept. of Health & Env.
Sciences
Solid & Hazardous Waste
Bureau
Cogswell Bldg.
Helena, MT 59620
(406) 444-2821

Nebraska

Waste Recovery Section
Dept. of Env. Control
P.O. Box 98922, State
House Sta.
Lincoln, NE 68509-8922
(402) 471-4210

Nevada

Waste Management Div.
Dept. of Conservation &
Natural Resources
123 W. Nye Lane, Rm. 142
Carson City, NV 89710
(702) 687-5872

New Hampshire

Dept. of Env. Services
Waste Management Div.
Health & Welfare Bldg.
6 Hazen Dr.
Concord, NH 03301-6509
(603) 271-2900

New Jersey

Solid Waste Admin.
Dept. of Env. Protection
840 Bear Tavern Rd.
CN 414
Trenton, NJ 08625
(609) 530-8591

New Mexico

Solid Waste Program
Env. Improvement Div.
1190 St. Francis Dr.
Santa Fe, NM 87503
(505) 827-2780

New York

Div. of Solid Waste
Dept. of Env. Conservation
50 Wolf Rd.
Albany, NY 12233-4010
(518) 457-6603

North Carolina

Dept. of Env., Health &
    Natural Resources
Div. of Solid Waste
    Management
Hazardous Waste Section/
    Solid Waste Section/
    Superfund Section
P.O. Box 27687
Raleigh, NC 27611-7687
(919) 733-2178 - Hazardous
    Waste
(919) 733-0692 - Solid Waste
(919) 733-4996 - Div. of Solid
    Waste Management

North Dakota

Div. of Waste Management
Dept. of Health
P.O. Box 5520
Bismarck, ND 58502-5520
(701) 221-5166

Ohio

Env. Protection Agency
Div. of Hazardous Waste
    Management
1800 Watermark Dr.
Columbus, OH 43266
(614) 644-2917

Div. of Solid & Infectious
    Waste Management
1800 Watermark Dr.
Columbus, OH 43266
(614) 644-3181

Oklahoma

Solid Waste Management Sec.
Dept. of Health
1000 N.E. 10th St.
Oklahoma City, OK 73117-1299
(405) 271-7159

Oregon

Solid Waste Management Div.
Dept. of Env. Quality
811 S.W. 6th Ave.
Portland, OR 97204
(503) 229-5913

Pennsylvania

Bureau of Waste Management
Dept. of Env. Resources
Fulton Bldg., 8th Floor
P.O. Box 2063
Harrisburg, PA 17105-2063
(717) 787-9870

Rhode Island

Dept. of Env. Management
Div. of Air & Hazardous
  Materials
291 Promenade St.
Providence, RI 02908
(401) 277-2797

South Carolina

Health Services
Dept. of Health & Env. Control
1751 Calhoun St., Mills
  Complex
Rm. 0106
Columbia, SC 29201
(803) 737-4180

South Dakota

Div. of Env. Regulations
Dept. of Water & Natural
  Resources
Joe Foss Bldg., 523 E. Capital
Rm 223
Pierre, SD 57501
(605) 773-3153

Tennessee

Div. of Solid Waste
  Management
Bureau of Env. Services
Customs House
701 Broadway
Nashville, TN 37247-3530
(615) 741-3424

Texas

Bureau of Solid Waste
  Management
Texas Dept. of Health
1100 W. 49th St.
Austin, TX 78756
(512) 458-7271

Utah

Solid & Hazardous Waste
  Bureau
288 North, 1460 West
P.O. Box 16690
Salt Lake City, UT 84116
(801) 536-6170

Vermont

Agency of Natural Resources
Solid Materials Div.
103 S. Main
Waterbury, VT 05676
(802) 244-7831

Virginia

Dept. of Waste Management
101 N. 14th St.
Monroe Bldg., 11th Floor
Richmond, VA 23219
(804) 225-2667

Washington

Dept. of Ecology
Solid Waste Program
Mail Stop PV-11
Olympia, WA 98504-8711
(206) 438-7440

West Virginia

Solid Waste Office
1456 Hansford
Charleston, WV 25301
(304) 348-5993

Wisconsin

Bureau of Solid Waste
  Management
Dept. of Natural Resources
P.O. Box 7921
Madison, WI 53707
(608) 266-2111

Wyoming

Solid Waste Management
  Program
Dept. of Env. Quality
Hershler Bldg.
122 W. 25th St.
Cheyenne, WY 82002
(307) 777-7752

Radiation

Alabama

Dept. of Env. Management
  Air Div.
1751 Congressman W.L.
  Dickinson Dr.
Montgomery, AL 36109
(205) 271-7700

Alaska

Radiation Control Program
Dept. of Env. Conservation
P.O. Box O
Juneau, AK 99811-1800
(907) 465-2666

Arizona

Radiation Regulatory Agency
4814 S. 40th St.
Phoenix, AZ 85040
(602) 255-4845

Arkansas

Radiation Control Program
Dept. of Health
4815 W. Markham
Little Rock, AR 72205-3867
(501) 661-2301

California

Dept. of Health Services
Radiological Health Branch
1231 Q St.
Sacramento, CA 95814
(916) 445-0931

Colorado

Radiation Control Div.
Dept. of Health
4210 E. 11th Ave.
Denver, CO 80220
(303) 331-8480

Connecticut

Bureau of Air Management
Radiation & Noise Control Div.
165 Capital Ave.
Hartford, CT 06106
(203) 566-5668

Delaware

Dept. of Public Health
Bureau of Env. Health
PO Box 637, Cooper Bldg.
Dover, DE 19903
(302) 739-4731

Florida

Office of Radiation Control
Dept. of Health
Rehabilitation Services
HRS/HSERR
1317 Winewood Blvd.
Tallahassee, FL 32399-0700
(904) 488-1525 Radon Control
(904) 487-1004 Radiation
   Control

Georgia

Env. Protection Div.
Dept. of Natural Resources
205 Butler St., S.E.
Atlanta, GA 30334
(404) 656-6905

Hawaii

Noise & Radiation Branch
Dept. of Health
591 Ala Moana Blvd.
Honolulu, HI 96813
(808) 548-3075

Idaho

Div. of Env. Quality
Hazardous Materials Bureau
Dept. of Health & Welfare
1410 N. Hilton St.
Boise, ID 83706
(208) 334-5879

Illinois

Dept. of Nuclear Safety
1301 Knotts St.
Springfield, IL 62703
(217) 786-6384

Indiana

Div. of Industrial Hygiene &
   Radiological Health
Board of Health
1330 W. Michigan St.
Indianapolis, IN 46206
(317) 663-0147

Iowa

Dept. of Env. Quality
Henry A. Wallace Bldg.
900 E. Grand
Des Moines, IA 50319
(515) 281-8692

Kansas

Radiation Control Program
Bureau of Env. Health Services
Dept. of Health & Env.
Mills Bldg., Rm. 602
109 S.W. 9th St.
Topeka, KS 66612-1274
(913) 296-1560

Kentucky

Dept. of Health
Div. of Community Safety
275 E. Main St.
Frankfort, KY 40621
(502) 564-4537

Louisiana

Dept. of Env. Quality
Radiation Protection Div.
PO Box 14690
Baton Rouge, LA 70898
(504) 925-4518

Maine

Radiological Health Program
Div. of Health Engineering
Dept. of Human Services
State House, Station 10
Augusta, ME 04333-0100
(207) 289-5676

Maryland

Dept. of Env.
Center for Radiological Health
2500 Broening Hwy.
Baltimore, MD 21224
(301) 631-3000

Massachusetts

Air Management Div.
Env. Protection Agency
JFK Federal Bldg.
Apt-III
Boston, MA 02203
(617) 565-3234

Michigan

Div. of Radiological Health
Dept. of Public Health
P.O. Box 30195
3423 N. Logan St.
Martin Luther King, Jr. Blvd.
Lansing, MI 48909
(517) 335-8190

Minnesota

Radiation Control
Dept. of Health
925 S.E. Delaware St.
P.O. Box 59040
Minneapolis, MN 55459
(612) 627-5014

Mississippi

Div. of Radiological Health
P.O. Box 1700
Jackson, MS 39215-1700
(601) 354-6657

Missouri

Bureau of Radiological Health
Dept. of Health
1730 E. Elm St., PO Box 570
Jefferson City, MO 65102
(314) 751-6083

Montana

Occupational Health Bureau
Cogswell Bldg., Capital Sta.
Helena, MT 59620
(406) 444-3671

Nebraska

Hazardous Waste Section
Land Quality Div.
Dept. of Env. Control
P.O. Box 98922
State House Sta.
Lincoln, NE 68509-8922
(402) 471-2186

Nevada

Dept. of Health
Radiological Health Section
505 King St., East, Rm. 203
Carson City, NV 89710
(702) 687-5394

New Hampshire

Div. of Public Health Services
Bureau of Radiological Health
6 Hazen Dr.
Concord, NH 03301-6527
(603) 271-4588

New Jersey

Radiation Protection Program
CN-415
Trenton, NJ 08625-0415
(609) 987-6393

New Mexico

Env. Improvement Div.
Radiation Licensing &
   Registration Section
1190 St. Francis Dr.
Santa Fe, NM 87503
(505) 827-2948

New York

Hazardous Substances
   Regulation
Div. of Env. Conservation
Bureau of Radiation
50 Wolf Rd., Rm. 510
Albany, NY 12233-7255
(518) 457-2225

North Carolina

Dept. of Env. Health & Natural
   Resources
Div. of Radiation Protection
P.O. Box 27687
Raleigh, NC 27611-7687
(919) 733-4283

North Dakota

Dept. of Health
Div. of Env. Engineering
1200 Missouri Ave., Rm. 304
Bismarck, ND 58502-5520
(701) 221-5188